The Measurement and Analysis of Housing Preference and Choice

Sylvia J.T. Jansen • Henny C.C.H. Coolen
Roland W. Goetgeluk
Editors

The Measurement and Analysis of Housing Preference and Choice

Springer

Editors
Sylvia J.T. Jansen
Delft University of Technology
OTB Research Institute
for the Built Environment
Jaffalaan 9
2628 BX Delft
Netherlands
s.j.t.jansen@tudelft.nl

Henny C.C.H. Coolen
Delft University of Technology
OTB Research Institute
for the Built Environment
Jaffalaan 9
2628 BX Delft
Netherlands
h.c.c.h.coolen@tudelft.nl

Roland W. Goetgeluk
Demography & Housing
ABF Research, Delft
The Netherlands
roland.goetgeluk@abf.nl

ISBN 978-90-481-8893-2 e-ISBN 978-90-481-8894-9
DOI 10.1007/978-90-481-8894-9
Springer Dordrecht Heidelberg London New York

Library of Congress Control Number: 2011928082

Printed on acid-free paper

Springer is part of Springer Science+Business Media (www.springer.com)

Contents

Chapter 1
Introduction

Sylvia J.T. Jansen, Henny C.C.H. Coolen, and Roland W. Goetgeluk

1.1 Aim and Perspective

This book contains an overview of methods and analytical techniques that can be used to describe, predict, and explain housing preference and housing choice. Its purpose is to make the choice of a method or technique to answer a specific research question easier. The book is meant for stakeholders involved in housing studies and practice, including students and researchers in housing research, marketing, business economics, management, business administration, spatial planning, and human geography. We hope that the book proves to be a useful resource for academics that have an interest in the measurement of housing preferences. Furthermore, we also focus on professionals in practice, such as policy-makers, landlords, property developers, and constructors. To them, this book may serve as a guide that gives insight into the differences between the research methods as well as insight into the interpretation of results obtained with a particular method.

A key factor in research is the method: it defines the results. A mutual understanding of the basics of various methods is a necessary condition for supporting research. To select a valid research method, one needs a well-structured overview of the methods and techniques commonly available in housing preference research. However, there is no comprehensive introduction to this field. This book aims to fill this gap and offers such an overview. The selection of methods is based on our experiences in fundamental and applied research as well as in education. Providing information, comparing the characteristics and describing the potential limitations of each method is an important instrument in order to be able to make a choice from various alternatives. This is the very core of this book. We have attempted to provide

S.J.T. Jansen (✉), H.C.C.H. Coolen
OTB Research Institute for the Built Environment, Delft University of Technology, Delft, The Netherlands
e-mail: s.j.t.jansen@tudelft.nl; h.c.c.h.coolen@tudelft.nl

R.W. Goetgeluk
Demography & Housing, ABF Research, Delft, The Netherlands
e-mail: roland.goetgeluk@abf.nl

S.J.T. Jansen et al. (eds.), *The Measurement and Analysis of Housing Preference and Choice*, DOI 10.1007/978-90-481-8894-9_1,

a broader perspective than merely a methodological one. The emphasis lies on the descriptions of methods and analytical techniques related to the practical framework of goals in housing studies, in particular "why do people move?," "what do customers want?," and "which choices do they make?."

1.2 Distinction Between Housing Preference and Housing Choice

Although the concepts of preference and choice are widely used in housing research these terms are sometimes mistaken for each other. In our view, preference refers to the relative attractiveness of an object, while choice refers to actual behavior. Preference, as an expression of attractiveness, may guide choice, but the evaluation involved in preference may take place whether or not a choice has to be made. What complicates the distinction between preference and choice is the notion that hypothetical choices, as can be made in, for example, the Conjoint Analysis method, should be seen as an expression of preference and not of choice. Thus, if a hypothetical choice is made in favor of dwelling A instead of B, this indicates a preference for dwelling A over B. As we will describe later in this chapter, the methods described in this book cover the whole range from housing preferences to actually made housing choices.

The most important difference between housing preference and housing choice is that preference is a relatively unconstrained evaluation of attractiveness. In the case of a house, choice will always reflect the joint influences of preference, market conditions, regulations, availability, and internal and external personal factors such as lifestyle and social class. Housing preference might not show a strong relationship with the housing choice actually made.

Priemus (1984) argues that numerous factors limit the number of realistic possibilities for every household, such as:

- Government regulations.
- Supply factors (such as place, nature, and price of the available dwellings).
- Preferences of the household.
- Transparency of the housing market (how well does the household know the supply side).
- The budget of the household (determined by financial position, income, and income perspectives).

This list can be extended with factors that influence consumer behavior, such as information-seeking behavior, time constraints, lifestyle, culture, family, motivation, social class, reference groups, and perception (Gibler and Nelson 2003).

These factors result in restrictions that limit the household's choice of a dwelling. Therefore, often there is not so much choice, although choice becomes greater when the household has a larger budget. Because of these factors, actual behavior (so-called revealed preferences) often differs substantially from their original preferences (so-called stated preferences).

Hooimeijer (1994) describes that the motivating power for adjusting housing preferences occurs at two, mutually related, levels. The first level is that of the (regional) housing market and the second one the level of the individual households. Households' preferences change through changes in their individual life course and their housing market position and because the composition of the supply changes (the occurrence of new practical or theoretical opportunities) (Hooimeijer 1994).

1.3 Conceptual Frameworks for Studying Housing Preference and Choice

Housing preference and choice has been studied from different theoretical perspectives. A detailed description of all these methods and approaches lies beyond the scope of this book. The interested reader is referred to, for example, Sabagh et al. (1969), Ritchey (1976), Priemus (1984), Fawcett (1986), Musterd (1989), Smid and Priemus (1994), Timmermans et al. (1994) and Mulder (1996). Below we will describe some of the available theories in the field of housing preference and choice. The first, the life-cycle and life-course models, explain and predict residential mobility. The second, the Theory of Planned Behavior, is a general theory to model behavior based on attitude, social norm, and perceived behavioral control. The third is a model of decision-making, applied to the specific case of housing. For reasons of clarity, the models are described as being separate models. However, they have mutual relationships and have and have had influence on each other, now and in the past. Furthermore, the short description of the models does not claim to be exhaustive.

1.3.1 Life-Cycle and Life-Course Models

The family life-cycle model (Rossi 1955) and its more recent adaptation and extension, the life-course model, is one of the most important models to explain housing moves. According to the original family life-cycle model, different stages of nuclear family formation (cohabitation/marriage), expansion (birth of children), contraction (children moving out), and dissolution (divorce or death of a spouse) lead to changes in the size and composition of households as well as in their residential preferences and needs. A transition into a new stage in the cycle may lead to a mismatch because housing characteristics, such as the number of bedrooms, might no longer meet the needs or preferences of the family. This leads to dissatisfaction (complaints), which emerges as a central motivational construct in Rossi's analysis (Fawcett 1986). For example, the birth of the first child, thus the transition of couple to family, is frequently preceded or followed closely in time by a move from renting to owning (Clark et al. 1994). Bell (1958) argues that the move to the suburbs that he observed in families with children can be explained as an effort to find a place for family life that is more suitable than that offered by central cities.

In contrast to the life-cycle model, which represents family formation and development as a fixed sequence of static states, the life-course model analyzes the sequence of positions of a particular person or group in the course of time. It is a longitudinal approach. Furthermore, the interaction between individual lives and social change is studied (Kok 2007). In a life-course analysis, the frequencies and timing of changes in positions, such as leaving home, is studied. These changes in positions are called events or transitions (Kok 2007). Every life course consists of a sequence and a combination of transitions, these are called trajectories. The time between transitions is called duration (Kok 2007). In trying to understand life courses, the life-course analysis makes use of both quantitative (structured) data and qualitative (less-structured) data, such as interviews, biographies, letters, and data from other sources.

In the life-course perspective, people are regarded as following parallel, intertwining careers in different life spheres: a labor market career, a housing career, etc. (Mulder and Hooimeijer 1995). People try to combine goals that arise from various careers in time and therefore in space. In housing choice research, four careers may explain moves: the educational, the labor, the family, and the housing or residential career. Clark et al. (2003) explain that the housing career is the sequencing of housing states defined in terms of tenure and the quality/price of the dwellings that households occupy. Goetgeluk and Hooimeijer (1991), Goetgeluk et al. (1992) and Mulder (1993) showed that each of the four careers might be the triggering career for a move, while the others form the conditional careers. For example, starting a university career implies an induced move if commuting is impossible (Clark and Onaka 1983). A student's housing choice is often not equal to a "maximum utility" from the housing career perspective. Relocation is a strategic spatial and costly choice to combine all careers of the members of the households in the short- and the long-term. Goetgeluk and Hooimeijer (1991), Goetgeluk et al. (1992) and Mulder (1993) showed how the four motives for moving influenced the urgency to move, the search-time, and therefore the knowledge of the local market, the number of dwelling attributes that were considered, the urgency to accept and the probability of moving. They also showed how the conditional careers influence choice and that the different careers interact. Having financial resources from work might imply living in a luxurious dwelling; economizing on the costs of housing might result in spending money on other purposes, such as a holiday trip. Life-course analysis also studies the interaction between life course and demographic, economic, institutional, and social changes (Kok 2007). Thus, the framework does take account of external conditions like market conditions (demand–supply) and allocation rules. For instance, a high income precludes a household from entering the social rented sector. Goetgeluk (1997) showed that the specific conditions of a housing market region influenced the preference and choice structure of demand.

1.3.2 *The Expectancy-Value Model and the Theory of Planned Behavior*

Expectancy-Value theory was developed in order to explain and predict the attitude toward objects and actions. Behavior, behavioral intentions or attitudes are seen as a function of (1) expectancy, i.e. the perceived probability that an object possesses a particular attribute or that a behavior will have a particular consequence, and (2) value, i.e. the degree of affect, positive or negative, toward an attribute or behavioral consequence. The model proposes that an attitude is a function of the sum of the expected values of the attributes. In the late 1970s and early 1980s, Fishbein and Ajzen expanded the Expectancy-Value theory into the Theory of Reasoned Action (TRA; Fishbein and Ajzen 1975). Later, Ajzen posited the Theory of Planned Behavior (TPB; Ajzen 1991). According to TPB, behavior is guided by three types of considerations (see Fig. 1.1). The first type, attitude, concerns beliefs about the perceived consequences of the intended behavior (behavioral beliefs). The perceived likelihood of the positive and negative consequences of the behavior is combined with the evaluation of these consequences to form the attitude. This part of the model resembles the Expectancy-Value model described earlier. The second consideration, subjective norm, is based on normative beliefs that reflect a person's perception of what referent individuals or groups think that he or she should do, combined with the person's motivation to comply with these referents. The third consideration, the perceived behavioral control, reflects the person's perceived ease or difficulty of performing the behavior considering the potential barriers and opportunities. It incorporates beliefs about the presence of factors that may facilitate or impede the behavior, combined with the perceived possibilities of having control over these factors (control beliefs). According to TPB, attitude, subjective norm, and

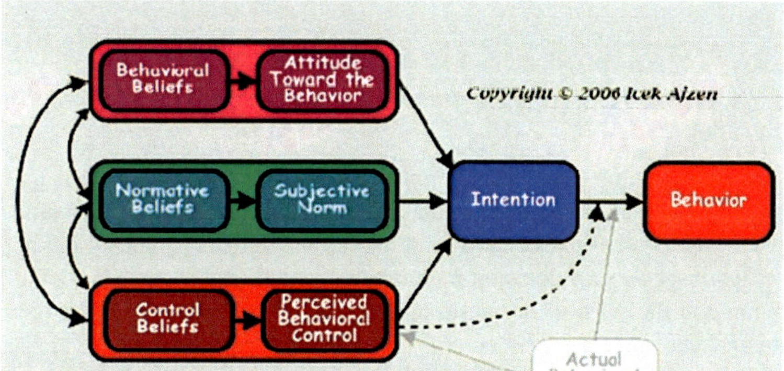

Fig. 1.1 Theory of planned behavior (Source: http://people.umass.edu/aizen/tpb.diag.html)

perceived behavioral control work together to form the behavioral intention to perform the specific action. The intention may result in actually performing the behavior but this is dependent upon actual behavioral control. The person must have a sufficient degree of actual control over the behavior in order to be able to act.

An Expectancy-Value model of migration decision-making was introduced in housing research by de Jong and Fawcett (1981) (see also: de Jong et al. 1983, 1986). Ratings of importance (the values) are obtained for the set of relevant values or goals. Next, measures of expectations, i.e. the probabilities of achieving these values or goals in specific locations, are obtained. Finally, a formula is applied to derive a place-specific attraction score, which is used as a predictor of migration intentions or behavior (De Jong et al. 1983). A relatively high summation score for a particular location is posited to indicate a propensity to move to that place. De Jong et al. (1983) explored whether expectations of attaining important economic and noneconomic values or goals were determinants of the decision to stay within their province, to move outside their province but stay within their country (the Philippines) or to move outside the Philippines. In addition, the authors explored whether there were other factors, next to value-expectations, that had an impact on the intention to move, such as personality traits, risk-taking orientation, and household characteristics. They also explored perceptions of social norms about migration, resembling the subjective norm part of TPB. Interestingly, de Jong et al. (1986) refute the Ajzen and Fishbein (1980) position that intentions are the dominant determinant of behavior. De Jong et al. (1986) argue that personal and structural background factors also exert an independent and direct effect on migration intention and behavior. In their study, two of the most important predictors of the general intention to move or stay were being single and having money to move. Other studies that have implemented the Expectancy-Value approach in housing are Fokkema (1996) and Fuller et al. (1986).

1.3.3 Decision-Making Approach

Another approach to unraveling the underlying considerations of housing choice comes from the domain of decision-making. Whereas the Expectancy-Value model focuses on the content of decisions and intentions, the Decision-Making approach focuses on the process of decision-making: how people make choices about moving or staying and the selection of destination. Here, moving house, or the possibility of moving, can be seen as a complex and ill-structured problem which can be solved by means of human problem-solving techniques (Rossi 1955; Newell and Simon 1972; Holland et al. 1986; Simon et al. 1987). Searching for a new dwelling can be seen as a dynamic problem-solving process in which a relatively unfamiliar, complex, and ill-structured problem must be solved. By this, we mean that people who are looking for another house generally do not have such well-articulated preferences. We believe that a housing preference is, at least partly, constructed during the problem-solving process. As Gregory et al. (1993) argue, the forming of preferences may be more like architecture, building a defensible set of values, rather than archeology, uncovering values that are already there. The phenomenon that preferences are constructed in the process of solving problems by adaptive humans, especially when important, complex and unfamiliar problems are at stake, is relatively common in human problem-solving (Payne et al. 1992; Slovic 1995). Individuals decide how to decide by considering various goals, such as reducing the cognitive effort required for making a choice, maximizing the accuracy of the decision, minimizing the experience of negative emotion, and maximizing the ease of justifying the decision (Bettman et al. 2006, p. 325). Individuals try to find strategies that will yield high degrees of accuracy for reasonable amounts of effort in any given decision task. Often, however, individuals must make trade-offs between accuracy and effort in selecting a strategy. The relative accuracy and effort levels of various strategies change with task demands (Payne et al. 1992). The above implies that in decision-making, rationality of individuals is limited by the information they have, their cognitive limitations and the finite amount of time they have to make decisions. This is termed bounded rationality (Simon 1991). Therefore, decision-makers may not arrive at the optimal solution (from an economic viewpoint) but instead apply their rationality only after having greatly simplified the choices available.

Problem-solving is value-focused and goal-oriented behavior. People try to achieve certain goals and values when solving their problems. A goal implies a certain desired end state, such as freedom, happiness, and security. Problem-solving occurs in a series of analytically well-defined stages, although a particular problem could repeat and backtrack in a complex way. Several conceptualizations of the problem-solving process can be found in the literature on decision-making and on consumer choice (e.g., Simon 1977; Janis and Mann 1977; Huber 1986, 1989; Engel et al. 1995; Carroll and Johnson 1990; De Groot 1978; De Groot and Gobet 1996; Beach 1990). We distinguish the following stages of problem-solving: (1) recognition; (2) formulation; (3) designing and screening; (4) choice; (5) deliberating about commitment; (6) action; and (7) feedback. Below we provide a detailed description of these stages.

The process of problem-solving begins with the realization by a household that there are problematic conditions that call for a solution (Recognition). In the context of housing choice this means that the household is confronted with a problem for which moving to another housing unit is considered one of the solutions. The household may therefore experience the current housing situation as unsatisfactory or may expect it to become so in the near future. Or the household may consider the option of moving as instrumental to solving another problem or preference such as living closer to work or relatives, having more privacy, becoming unemployed, a divorce, and so on.

When a situation is recognized as an issue calling for a solution, the next stage (Formulation) involves exploring and classifying the situation, including some understanding of relevant objectives (based on preferences) and values. In this stage, the household constructs an internal representation (frame) of the problem situation.

This frame will also contain some ideas about the goals and values (e.g., more space, quiet neighborhood, more privacy) to be achieved and the constraints (e.g., the amount of money to be spent) that a subjectively acceptable solution must meet.

After the household has formulated the problem situation, the attention is focused on one or more alternative solutions (Designing and screening). The household searches in memory for alternative courses of action and seeks advice and information from others about ways of coping with the problem. In the context of housing choice, this means that one tries to remember how the problem was handled in earlier moves, if this is not the first move. The household may read the housing advertisements in newspapers and other periodicals more attentively, ask friends and acquaintances what ideas they may have about solutions, make contact with realtors or project developers and so on. As the household screens one or more alternative solutions during this stage, any solution that appears not to meet the goals (based on preferences and constraints) is eliminated from further consideration. By the end of this stage, the household has narrowed down the set of alternative solutions to those that appear to have a good chance of solving the problem. These alternative solutions constitute the so-called consideration set (also known as the evoked set) (Engel et al. 1995).

In the subsequent stage of Choice, the household employs a more thorough search and evaluation, focusing on the positive and negative aspects of the remaining alternative solutions in order to select the best available course of action. By best course of action, we do not mean an optimal course. In general, we expect problem-solvers to search for a subjectively satisfying solution (De Groot and Gobet 1996). In an effort to select such a course of action people use a great variety of decision rules (Svenson 1979; Payne et al. 1993).

After having made a decision, the household begins to deliberate about implementing it and conveying this intention to others (Deliberating about commitment). The household becomes concerned about the possible disapproval of others. These fresh concerns may deter the household from taking immediate action without first paving the way by letting others know about the intended choice. As the household approaches the point of implementing the decision, it also realizes that once a decision is made it is more difficult to reverse it. This realization makes for reconsideration of the choice made. For instance, the high cost involved in moving may lead to some reflection.

Given the fact that the household wants to bolster and consolidate the decision in a way that will enable the household to implement it, the decision must still be acted upon (Action). The fact that the household wants to implement the solution does not mean that it will also be implemented. The chosen dwelling may no longer be available for instance, or the necessary financing for the housing unit cannot be found.

After the decision has been acted upon, the household receives information about the outcomes of the action (Feedback). This feedback may be both positive and negative. It enables learning about the substance of the solution and about the decision rules employed. The household also receives feedback when it justifies its decisions to others, so this phase may stretch out over a long period of time. If the public services (Dieleman et al. 1989). Differences in location might influence various aspects, such as social status, the consumption of private goods, the availability of public goods, jobs, and other desired destinations. Location is therefore an important determinant of housing quality and household welfare.

Thirdly, a high structural durability means that an overwhelming share of current dwellings is provided by units built in the past. Housing is expensive to build, modified only with difficulty, and one of the most lasting of consumer durables (Clark and Dieleman 1996). Newly produced units contribute a relatively small percentage to the total stock. The supply of dwellings, both in numbers and in quality, is inflexible in the short term. This may create shortages in periods of rapid growth of urban populations and considerable fluctuations in rents and prices (Clark and Dieleman 1996).

Fourthly, nonetheless, the existing stock is continuously undergoing modification in various ways to accommodate the changing latent demand. New housing units are added to the housing stock and units may be lost through demolition, fire, conversion, or abandonment (Bourne 1981). Quality, condition, structural features, and size change over time, as do the number of separate dwelling units in a structure and the occupant's tenure status. Supply does not take the form of modules of a homogeneous good. On the contrary, housing units are enormously heterogeneous. They differ in numerous structural characteristics, lot features, neighborhood characteristics, local public services, and access to desired destinations. Each unit differs, if only slightly, in design and location (Bourne 1981). Housing is a package or bundle of many salient attributes, only some of which are under the control of the resident.

The former has made clear that housing is firmly embedded in the social, economic, and political aspects of society. Bourne (1981) makes a distinction between eight areas of housing research. These areas vary in scale (micro, macro) and in subject matter (demand, supply, policy). On the demand side, he distinguishes literature on household location decisions, residential mobility, land use and neighborhood change, social and demographic change and political and economic structures. On the supply side, the relevant areas include studies in investment and capital markets, institutional behavior, and local development practices. The enormous breadth of the literature on housing necessitates the choice for a confined research topic. Therefore, the current book limits itself to providing a description and explanation of nine different methods and analytical techniques for measuring housing preference and choice. We acknowledge, however, that housing preference and choice do not materialize in isolation.

1.5 Different Approaches

Now that we have described some of the theoretical background of measuring housing preference and choice as well as the special features of housing, it is time to introduce the nine methods and analytical techniques that form the core of this book.

The issues of housing choice and housing preference attract interest from researchers in a variety of disciplines. Because of this multidisciplinarity, research into housing preference gives rise to numerous different approaches and models. For example, economists primarily focus on observed house prices to determine the utility of dwelling characteristics (called attributes). In contrast, social geographers are mostly concerned with the influence of socio-demographic changes on residential moves.

What consumers want can be measured in many different ways. Which particular method is to be chosen can only be answered in the light of the purpose of the measurement (Hooimeijer 1994). Different methods lead to different outcomes. The choice for a specific method can therefore not be based on the methodological superiority of one method over another but should be directed by the type of information in which one is interested (Hooimeijer 1994). The selection of a particular method can be based, for example, on the desired outcome measure or on the source of the preferences (stated or revealed preferences). Based on our experience with research and education in the field of housing studies, we selected nine methods and analytical techniques that are currently applied in the research field. Each of these methods will be briefly described.

Note that we do not have the intention to present all possible or even all relevant methods and techniques that are available in the research field of housing preference and choice. We have restricted ourselves to methods that have been frequently used and for which a description of the method as well as an example in practice in the domain of housing preference research was available.

1.5.1 Traditional Housing Demand Research Method

Probably the most well-known method for eliciting housing preference is the traditional housing demand research method. In general, relatively simple and straightforward questions are asked about the willingness to move, preferences for housing (environment) characteristics and the current and previous (in the case of a recent move) housing situation. Furthermore, socio-demographic and economical variables are collected, such as type of household and income. The results are usually analyzed and reported in a straightforward way. The goal of the traditional housing demand research method is to obtain accurate insight into the current and future demand for housing, in a quantitative as well as in a qualitative sense.

Often, these cross-sectional studies are repeated in time, but with another respondent sample. An analysis can be performed on the data obtained with such repeated cross-sectional studies, in order to examine whether preferences or choices change over time. An example of such a study is reported in the chapter on the traditional housing demand research method. In this study, the researcher examined the influence of demographic, socioeconomic, and sociocultural factors on the demand for (semi-)detached owner-occupied dwellings in the Netherlands in the period 1975–1997.

1.5.2 Decision Plan Nets Method

A Decision Plan Net can best be described as the underlying protocol that people use to evaluate alternative houses in terms of the housing attributes that are important to them. The purpose of the method is to uncover this underlying protocol and to represent it in a tree or flow diagram. This is done by first recording for each important housing attribute the individual's preferred level. These preferred housing attribute levels are represented on the main axis of the tree diagram. Subsequently, the importance of each preferred housing attribute level is determined. This means that respondents are asked to indicate the levels at which they would no longer consider that choice alternative to be acceptable, keeping all other housing attributes at the desired level. If the individual indicates that a house without this attribute level would be refused, the attribute is called a reject-inducing attribute, and this is represented in the tree diagram. The attribute is called a trade-off attribute if the individual indicates that a different level of the attribute in question will be accepted if this is compensated for by better scores on one or more of the other housing attributes. This is subsequently recorded in the tree diagram together with the compensatory attribute levels. Finally, the individual may indicate that a different level on the attribute in question would be accepted as long as the house satisfies the other housing attribute levels; such an attribute is called a relative preference attribute, and is also represented in the tree diagram. The final tree diagram clearly shows which dwellings are unacceptable and which housing alternatives are acceptable for the individual. By having the individual rank order these alternatives, one gets a clear picture of the individual's housing preferences.

1.5.3 Meaning Structure Method

The Meaning Structure method was introduced in housing research by Coolen and Hoekstra (2001). The purpose of the Meaning Structure method is to assess what people's housing preferences are and why they have these preferences. The Meaning Structure method relates houses and consumers. Each housing attribute is assumed to yield consequences, while the importance of consequences is based on their ability to satisfy people's personally motivating values and goals. A meaning structure chain relates the preference for a housing attribute to its contribution to the realization of objectives and values. A simple meaning structure chain consists of the trio: housing attribute level; consequence; value. This approach thus uncovers both people's preferences for housing attributes as well as their motives for these preferences. In this respect, the Meaning Structure method differs from most of the other approaches presented in this book that focus mainly on what people want. After the preferred housing attribute levels have been determined, people have to indicate for each housing attribute why the preferred level is important to them or what the preferred level means to them. This "why" question is repeated as a reaction to the answer to the first "why" question. The process stops when the individual can no longer answer the "why" question, or after a predetermined number of "why" questions, depending on the goal of the research and the research design. The aggregation of individual meaning structure chains and the subsequent analysis of this network into a meaning network may be done either by hierarchical value maps or by network analysis.

1.5.4 Multi-Attribute Utility Method

The Multi-Attribute Utility method was originally designed for situations in which a decision-maker has to make an optimal choice among a set of complex alternatives while taking all the facets, criteria, or attributes of each of the alternatives into account. The Multi-Attribute Utility method amounts to valuing and weighting each of the attributes, and subsequently combining the weighted values into an overall utility per alternative. The alternative with the highest utility represents the optimal choice. Although the Multi-Attribute Utility method was intended for choice situations, it can also be used for preference measurements. Given the salient housing attributes, it involves the following procedure. First, the levels of each housing attribute have to be transformed into numerical values by each individual involved in the research. A variety of methods is available for performing this task. A relatively easy method is to have individuals rate the levels of each attribute on a scale with two anchors: one extreme point of the scale – "not attractive at all"– is assigned the value 0 and the other end point – "extremely attractive" – is given the scale value 100. This results in individual scale values for all attribute levels. Next, individuals have to assign an importance weight to each housing attribute. Several methods are also available for this purpose. For instance, one might ask individuals to rate each housing attribute on a numerical scale with end point 0 meaning "not important at all" and 100 indicating "extremely important." The importance weights are normalized per individual in such a way that their sum is 1. Now, for each alternative an overall utility score can be computed. The most commonly applied aggregation rule to combine the values with the weights is the weighted linear model. It implies that the value of each attribute is multiplied by the attribute weight. Subsequently, these weighted scores are summed over all attributes of a particular alternative, resulting in a multi-attribute utility. This is done for all alternatives. A higher multi-attribute utility indicates a higher preference.

1.5.5 Conjoint Analysis Method

Conjoint Analysis is based on responses to residential profiles that are complete descriptions of the characteristics of the house and the housing environment, called attributes. The construction of these profiles is based on statistical designs that vary the residential attributes in a systematic way across all profiles. Residential profiles are usually included in rating or choice tasks. Rating tasks involve subjects, either individuals or complete households, expressing the strength of their preference for each profile separately on some rating scale, for instance, on a scale which runs from 0 (extremely unattractive) to 10 (extremely attractive). Choice tasks involve subjects choosing the most preferred residential profile out of a set of two or more different alternatives. This task is repeated for a specific number of predefined choice sets. In order to arrive at an overall rating or choice, subjects have to make trade-offs between the presented attributes. Hence, Conjoint Analysis is especially useful if one is interested in the trade-offs people make between residential attributes, e.g., in estimating the willingness to pay for certain qualities of a residence. To that effect, a utility model is estimated based on the observed responses that form the dependent variables while the varied attributes form the independent variables. When ratings are observed, a regression model is typically estimated, whereas a multinomial logit model is estimated if choices are observed. The result in both cases is a series of parameters that indicate to what extent each attribute level contributes to the overall utility that is derived from a residential alternative. This estimated utility function provides insight into the importance of each residential attribute and the trade-offs made among the attributes. The utility model allows one to predict the overall utility for any residence described in the included attributes. Furthermore, if choice tasks are applied, it allows one to predict the probability that each alternative out of a set of residential alternatives will be chosen. To summarize, the aim of Conjoint Analysis is to estimate utility functions that can be used to compare residential alternatives in terms of peoples' preferences.

1.5.6 Residential Images Method

The research method called Residential Images method ("Woonbeeldenonderzoek") tries to get nearer to the realistic house-hunting process by showing a catalog of available or potentially available prototypes of either existing or newly built housing. Pictures or drawings of dwellings are shown, usually accompanied by written information about characteristics that are difficult to show on pictures, e.g., the number of rooms. Images can be included in research using one of two approaches. The first one is to confront respondents with objective information about new options on the supply side which they might not have known before. This can be used for developing new submarkets or niche markets like housing for singles, communities of senior citizens and housing in water-rich environments. Here the advantage of showing relatively new housing options to those who would otherwise probably have preferred familiar options is obvious. What is measured in these surveys must modestly be described as the degree of acceptance or rejection of new housing options at first sight. This approach of Residential Images does not provide information about the preference function that shows how each part-worth utility of an attribute contributes to the total utility of an image. However, considering the aim of the approach this is not problematic. In the second approach, the underlying preference functions are revealed that provide information on how new niche attributes add to the total utility of available alternatives. In this approach, an alternative is composed by means of systematically combining all attribute levels, as in the Conjoint Analysis method. The aim of this method is to force those on the demand side to trade off their individual wishes, like in a real house-hunting process, taking price and availability into consideration.

1.5.7 *Lifestyle Method*

In recent years, the Lifestyle method has obtained more and more importance in the field of housing preference research. Its advocates argue that, due to the increased heterogeneity in housing demand, the traditional indicators of social position, such as education and income, no longer suffice to explain and predict housing preferences. The Lifestyle method is supposed to fill this gap. Thus, it is assumed that people that fall within the same lifestyle category share the same housing preferences. Consequently, building or rebuilding can be performed in such a way as to attract specific lifestyle groups. There is a large variety in existing lifestyle classification systems. In general, researchers collect a large amount of data by having a number of respondents answer a large battery of questions on different topics (e.g., values and norms, interests, attitudes). Next, the researchers use analytical procedures, such as cluster analysis and correspondence analysis, to determine whether there are underlying patterns in the data that can be attributed to different lifestyle categories. Once this classification has a firm base, new respondents fill out a survey with a limited number of questions that have proven to distinguish well between respondents. Based on their answering patterns, respondents are allocated into specific lifestyle categories. This information can be used to build or adapt dwellings according to the type of residents that are expected to be going to live in the particular neighborhood. It can also be used as a marketing tool in order to approach the right consumer in the right way.

1.5.8 *Neoclassical Economic Analysis*

Central to the Neoclassical theory is the axiom of revealed preferences. It states that buyers and sellers are able to rank and value the bids and offers for goods on the market. The subjective value that households attach to a good gives rise to their bids. The exchange of goods (i.e. dwellings) only takes place among buyers who cannot find another seller who asks less and sellers who cannot find another buyer who bids more in a certain period. The optimal choices of sellers and buyers on the housing market can thus reveal their preferences for housing quality. The decision to exit or improve one's home, to choose a new home for relocation, to change from renting to owner-occupation and vice versa makes up the household's choice set. On the supply-side, the landlord buys, lets or sells property, the constructor demolishes, renovates and constructs dwellings and the developer acts as the intermediary. Aggregation of net construction (construction minus demolition), exit and relocation rates leads to housing supply and demand. The equality of demand and supply in turn yields the equilibrium house price and rent. In the Neoclassical approach on the housing market, the exit and relocation choice and all choices derived from them (e.g., duration of stay, shift from letting to owner-occupation) can be modeled as functions of the price or rent of the dwelling, the quality of the old or new dwelling and the location and the characteristics of the household. Aggregation of the exit and relocation rates over all households produces a well-defined house price or market rent equation. The equilibrium house price or market rent will be a function of the attributes of the dwelling and neighborhood alone. In the chapter concerning the Neoclassical approach, the Exit model and the Hedonic Price Analysis are described in more detail.

1.5.9 Longitudinal Analysis

With Longitudinal analysis the same sample of respondents is followed at different points in time. Longitudinal analysis can be performed in a number of ways using various statistical techniques. The goal of Longitudinal analysis is to examine how characteristics or circumstances at one point in time shape individual outcomes or decisions at a later point in time. To answer such research questions, one needs longitudinal data. One important topic in the domain of housing concerns the realization of intentions to move. The relationship between stated intentions and actual moving behavior can best be studied longitudinally following the same group of respondents in time. Such studies can provide insight into the extent to which people behave in accordance with their stated intentions to move and can provide insight into the circumstances that hamper or stimulate actual moving behavior. In the chapter on longitudinal research such a panel study is performed. The statistical technique of logistic regression analysis is used in this study to examine the impact of housing preferences, urgency to move, and characteristics of the respondent, the dwelling and the search location on actual moves.

1.5.10 The Goals of the Nine Methods and Analytical Techniques

The above-described nine methods and analytical techniques are summarized in Table 1.1. The table also provides a short overview of the general goal of each specific method or technique for measuring housing preference and housing choice.

Table 1.1 Nine methods and analytical techniques for measuring housing preference and housing choice

Methods and analytical techniques	Goal
Traditional Housing Demand Research method	To obtain accurate insight into the current and future demand for housing, in a quantitative as well as in a qualitative sense
Decision Plan Nets method	To reveal people's choice process based on individual mixes of dwelling (environment) characteristics that are deemed essential, those that can be compensated for and those that are deemed irrelevant
Meaning Structures method	To assess what people's housing preferences are and why they have these preferences
Multi-Attribute Utility method	To make a rational choice between available alternatives based on the dwelling profile that yields the most utility
Conjoint Analysis method	To estimate a utility function that can be used to predict the overall utility of residential profiles and thus to compare residential alternatives in terms of peoples' preferences
Residential Images method	To examine preferences for new alternatives holistically
Lifestyle method	To build/restructure/distribute dwellings according to lifestyle group preferences
Neoclassical economic analysis	To rank and assess the preferences for alternatives
Longitudinal analysis	Analysis of a specific research question regarding the question how characteristics or circumstances at one point in time shape individual outcomes or decisions at a later point in time

1.5.11 The Outcomes of the Nine Methods and Analytical Techniques

Aside from the goal, the type of outcome of the various research methods is also of importance. What type of outcome is preferred depends upon the particular research question. In this respect, no method is better than the other is; they just provide different outcomes. It is up to the researcher to decide which type of outcome fits the goal of the particular study best. For each method, a short description is provided in Table 1.2.

The traditional housing demand research method yields a quantitative description of housing preferences and of the willingness to move. The Decision Plan Nets method provides a substitution interval that defines a ranked set of houses that the consumer would consider acceptable.

The Meaning Structure method relates housing attributes to underlying values. An outcome in terms of a utility refers to a numerical strength of preference. The more utility a certain alternative provides, the more it is preferred. Methods that have been designed specifically to obtain utilities are the Multi-Attribute Utility method and the Conjoint Analysis method. These methods are strongly related to

Table 1.2 Type of outcome of the nine selected methods and analytical techniques for measuring housing preferences

	Type of outcome
Traditional Housing Demand Research method	A quantitative description of housing preferences and of the willingness to move
Decision Plan Nets method	The substitution interval that defines a ranked set of houses that the consumer would consider acceptable
Meaning Structure method	An overview of the preferred attribute level per housing attribute and the meanings of these housing attribute levels
Multi-Attribute Utility method	A multi-attribute utility (strength of preference) for every alternative
Conjoint Analysis method	A utility function that describes to what extent each attribute level contributes to the overall utility of a residential alternative
Residential Images method	A ranking of new alternatives
Lifestyle method	An assignment into a particular lifestyle group
Neoclassical economic analysis	Monetary estimates of the willingness to pay for and equilibrium price of alternatives
Longitudinal analysis	An indication of the stability of one or more variables or the relationship between two or more variables over time

the classical microeconomic theory of utility maximization and rational behavior. Von Neumann and Morgenstern (1947) and Savage (1954) have formulated the axiomatics of this theory.

The Residential Images method usually has outcomes in the form of a ranking of new alternatives. The Lifestyle method mostly arranges respondents into subgroups according to certain characteristics that they share. It is assumed that these subgroups also share housing preferences. The Neoclassical Economic Analysis provides monetary estimates of the willingness to pay for and the equilibrium price of alternatives. The longitudinal analytical technique provides an indication of the stability of variables over time or of the relationship between two or more variables over time.

1.6 Three Dimensions

Now that we have introduced the nine methods and analytical techniques that form the core of this book, we describe three dimensions that can help to detect and explain differences and similarities between the selected methods. Note that the potential limitations that are related to the three dimensions are described in Chap. 11 of this book. The three dimensions apply to: (1) the origin of the data (stated or revealed), (2) freedom of attribute choice, and (3) compositional or decompositional approach. An overview is provided in Table 1.3.

Table 1.3 Overview of methods and analytical techniques with regard to the three dimensions

Applies to	Origin	Design	
Dimensions	Stated or revealed	Freedom of attribute choice	Compositional versus decompositional
Traditional Housing Demand Research method	Stated	No	Compositional
Decision Plan Nets method	Stated	Yes	Compositional
Meaning Structure method	Stated	Yes	Compositional
Multi-attribute Utility method	Stated	Yes	Compositional
Conjoint Analysis method	Stated	No	Decompositional
Residential Images method	Stated	No	Decompositional
Lifestyle method	Stated	No	NA
Neoclassical economic analysis	Both	No	NA
Longitudinal analysis	Both	No	NA

NA not applicable

1.6.1 First Dimension: Stated or Revealed Preferences

The first dimension relates to the origin of the data: does it concern choices that have actually been made in the "real world" (revealed preferences) or stated choices and preferences in response to survey questions (stated preferences)? The latter type of analysis combines observations of elicited preferences and hypothetical choices with assumptions about the underlying processes of preference formation to yield predictions. The first seven methods in Table 1.3 yield stated preferences.

In contrast, the revealed approach is based on observed actual housing choices in real markets and it is assumed to reflect people's preferences (Timmermans et al. 1994). The technique combines observations of realized choices with assumptions about underlying decision processes to yield predictions (Manski 1999). For instance, in hedonic price models, the price of the dwelling is regressed on the observed housing attributes. This provides an indication of the "worth" (the preference) of the various housing attributes. Advocates of this approach argue that it is only in the act of choice that people can reveal their preferences.

The Neoclassical Economic Analysis and the Longitudinal Analysis can be performed independent of the origin of the data. They can be performed on either revealed or stated preferences, or on a combination of the two approaches. An excellent example of the latter is provided in Chap. 10 in which data on stated preferences (intentions to move) are linked to data on actual moving behavior obtained from a register.

1.6.2 Second Dimension: Freedom of Attribute Choice

The second dimension on which the methods and techniques might differ is freedom of attribute choice for the respondent. A method that allows freedom of attribute choice can be applied (but not necessarily so) in such a way that respondents can choose their own salient attributes. Respondents can be left entirely free in their choice of attributes but they can also be provided with a list of preselected attributes to choose from. Usually, they can also add additional attributes to the preselected list, if that is deemed necessary. The Decision Plan Nets method, the Meaning Structure method, and the Multi-Attribute Utility method are approaches that allow freedom of attribute choice. With these methods, respondents can first be asked which dwelling attributes are important to them. Based on these attributes, further enquiries can be made into the trade-off between attributes (Decision Plan Nets method), the underlying motives (Meaning Structure method) and the evaluation and importance of attribute levels (Multi-Attribute Utility method). Note that freedom of attribute choice comes at a cost. Such data are usually collected by face-to-face or telephone interviews, which are relatively time-consuming and costly. The other methods, and analytical techniques, usually apply a preselected set of attributes and attribute levels. In such designs, there is no freedom of attribute choice for the respondents.

1.6.3 Third Dimension: Compositional Versus Decompositional Methods

The third dimension relates to whether the measurement method is attribute-based (compositional) or alternative-based (decompositional). A decompositional method starts with evaluating alternatives and decomposes these into separate attributes. To estimate the contributions of the attributes and attribute levels, statistical methods are applied. Parameters for the attributes are derived from the decision-maker's holistic evaluative responses to profile descriptions designed by the researcher. The Conjoint Analysis method and the Residential Images method are examples of the decompositional approach.

The compositional approach starts with single attributes and combines these into alternatives. Housing preferences are explored by recording separately and explicitly how people evaluate housing attributes. The importance of each attribute can be weighted and combined with the values, using some algebraic rule, to arrive at an overall evaluation. Note, however, that not all methods explicitly calculate an overall evaluation for each possible alternative. Methods that are based on the compositional approach are the Traditional Housing Demand Research method, the Decision Plan Nets method, the Meaning Structure method, and the Multi-Attribute Utility method.

The Neoclassical economic analysis is generally based on predicting some overall dependent variable, such as house price or the probability of moving, from several predictors, which usually refer to the characteristics of the dwelling, the dwelling environment, and the inhabitants. For example, the Hedonic Price Analysis (Neoclassical Economic Analysis) is used to determine the "worth" of certain dwelling characteristics (attributes) by statistically inferring it from the house price. Such a procedure could be termed a decompositional approach. However, this is less clear for the Longitudinal Analysis. Therefore, this dimension is deemed not to be applicable for these analytical techniques.

1.6.4 Compensatory Versus Non-compensatory Methods

There exists a discerning dimension that we did not include in Table 1.3. This concerns the difference between compensatory and non-compensatory methods. Compensatory decision-making implies that a low value on one attribute can be compensated for by a high value on one or more other attributes. Thus, the specific alternative may still obtain a high overall evaluation score despite a low value on one or more attributes. In contrast, a non-compensatory decision method implies that a highly valued attribute cannot make up for a weakly valued one. The valuation of an attribute above or below a certain preferred threshold must therefore lead to the rejection of an alternative. Consumers can use cut-offs to qualify products, such as setting a limit on the minimum number of rooms in a dwelling. They may no longer consider a specific housing alternative as appropriate if it does not meet the specific criterion. This may be in accordance with the way in which people decide in reality, for example, a dwelling without a garden may not be acceptable to a family with young children, irrespective of the size of the living room or the number of bedrooms.

Our reason for not including this dimension in Table 1.3 is that we believe that almost all methods can be compensatory or non-compensatory depending on the way in which the questions are framed or the analysis is performed. For example, in the Multi-Attribute Utility method a linear additive function can be used to describe compensatory decision strategies. This means that evaluations for separate attribute levels are simply added to obtain an overall utility for a particular dwelling. A low evaluation for a particular attribute level can be compensated by high evaluations on other attributes. However, a multiplicative function, which may approximate non-compensatory preference structures, can also be applied. This means that low evaluations can hardly be compensated for. Furthermore, for the less statistically sophisticated methods, whether or not some method is compensatory might be dependent upon whether the trade-off of preferences is questioned. If respondents are allowed to reject an alternative based on its level of functioning on one or more attributes, the method used is non-compensatory. If they were not allowed to reject alternatives, the method used is compensatory.

1.6.5 Combinations of Methods and Techniques

Up to now, we have presented the methods and techniques separately. However, the methods can be seen as complementary. For example, the methods that allow freedom of attribute choice are relatively time-consuming and costly. They can be deployed in a relatively small sample of respondents to obtain insight into the salient attributes (levels) for the particular study. These salient attributes (levels) can subsequently be used in a method that does not allow freedom of attribute choice and that can be used in larger samples because it is relatively cheap and quick.

Furthermore, the combination of compositional and decompositional methods in one measurement task is possible. For example, the task can be split up into two parts. In the first part, respondents are asked to evaluate the attributes and the attribute levels separately. In the next part, a conjoint analysis task is performed. The underlying goal here is to let the respondent grow accustomed to the attribute levels in order to make the conjoint analysis task easier. For an example, see Vriens (1997).

Another combination of methods is described in Chap. 10. Here, stated preferences (who want to move?) and revealed preferences (who have moved?) are combined to explore the factors that can predict an actual move in respondents that have previously indicated that they have the intention to move. Earnhart (2002) also describes a combined study of stated and revealed preference data. In this study, more insight was obtained into the factors driving housing decisions.

Lindberg et al. (1988) described a study in which the methods of Laddering (Meaning Structure method) and Multi-Attribute Utility were combined in order to predict both preference ratings and choices with regard to housing. Boumeester et al. (2008) published a report in which the methods of Decision Plan Nets, Meaning Structure method, and Conjoint Analysis method were combined in order to reveal respondents' preferences with regard to housing as well as to obtain insight into the flexibility of their preferences and in their underlying motives.

1.7 Conclusion of This Introductory Chapter

In this chapter, we have shown that the concept of housing is a complicated one. A house can have various functions and it is a heterogeneous product in a special market. This makes the measurement of housing preferences a complicated matter. However, it is also a matter of utmost importance as housing is one of the primary necessities of life. It is in the way that residents' preferences are incorporated into maintaining, planning, and building real estate and the built environment that ultimately provides the most benefit to those who take shelter in their homes.

The present chapter tried to sketch a picture of the methods that are currently available in housing research and to provide some information on their characteristics. In the last section of this chapter, we introduced three dimensions on which the nine selected methods and techniques differ with regard to the origin of the data and their design. In addition, in the last chapter of this book the potential limitations that are related to the three dimensions are described in more detail. In the chapters in between, the nine methods and analytical techniques will be introduced and discussed, each time providing an example from practice. The optimal method or technique to elicit housing preferences is dependent upon the goal of the specific study and the way in which the characteristics of the various methods fit this goal best. Hopefully, the information provided in this book will be helpful in selecting the optimal method or technique for the problem at hand or in judging the results of earlier studies.

References

Ajzen, I. (1991). The theory of planned behavior. *Organizational Behavior and Human Decision Processes, 50*(2), 179–211.

Ajzen, I., & Fishbein, M. (1980). *Understanding attitudes and predicting social behavior.* Englewood Cliffs, NJ: Prentice-Hall.

Beach, L. R. (1990). *Image theory: Decision making in personal and organizational contexts.* New York: Wiley.

Bell, W. (1958). Social choice, life styles and suburban residence. In W. F. Dobriner (Ed.), *The suburban community.* New York: Putnam's.

Bettman, J. R., Luce, M. F., & Payne, J. W. (2006). Constructive consumer choice processes. In S. Lichtenstein & P. Slovic (Eds.), *The construction of preference.* New York: Cambridge University Press.

Boumeester, H. J. F. M., Coolen, H. C. C. H., Dol, C. P., Goetgeluk, R. W., Jansen, S. J. T., Mariën, A. A. A., & Molin, E. (2008). *Module Consumentengedrag WoON 2006, Hoofdrapport.* Delft: Onderzoeksinstituut OTB.

Bourne, L. S. (1981). *The geography of housing.* London: Edward Arnold.

Carroll, J. S., & Johnson, E. J. (1990). *Decision research. A field guide.* Thousand Oaks, CA: Sage.

Clark, W. A. V., & Dieleman, F. M. (1996). *Households and housing; choice and outcomes in the housing market.* New Brunswick, NJ: CUPR Press.

Clark, W. A. V., & Onaka, J. (1983). A review and interpretation of reasons for moving. *Urban Studies, 20,* 47–58.

Clark, W. A. V., Deurloo, M. C., & Dieleman, F. M. (1994). Tenure changes in the context of microlevel family and macrolevel economic shifts. *Urban Studies, 31*(1), 137–154.

Clark, W. A. V., Deurloo, M. C., & Dieleman, F. M. (2003). Housing careers in the United States, 1968–93: Modelling the sequencing of housing states. *Urban Studies, 40,* 143–160.

Coolen, H., & Hoekstra, J. (2001). Values as determinants of preferences for housing attributes. *Journal of Housing and the Built Environment, 16,* 285–306.

de Groot, A. D. (1978). *Thought and choice in chess* (2nd ed.). The Hague: Mouton.

de Groot, A. D., & Gobet, F. (1996). *Perception and memory in chess.* Assen: Van Gorcum.

de Jong, G. F., & Fawcett, J. T. (1981). Motivations for migration: An assessment and a value expectancy research model. In G. F. De Jong & R. W. Gardner (Eds.), *Migration decision-making* (pp. 13–58). New York: Pergamon.

de Jong, G. F., Abad, R. G., Arnold, F., Carino, B. V., Fawcett, J. T., & Gardner, R. W. (1983). International and internal migration decision making: A value-expectancy based analytical framework of intentions to move from a rural Philippine province. *International Migration Review, 17*(3), 470–484.

de Jong, G., Root, B., Gardner, R., Fawcett, J., & Abad, R. (1986). Migration intentions and behavior: Decision making in a rural Philippine Province. *Population and Environment, 8*(1/2), 41–62.

Dieleman, F. M. (1996). Modeling housing choice. *Journal of Housing and the Built Environment, 11,* 201–207.

Dieleman, F. M., Clark, W. A. V., & Deurloo, M. C. (1989). A comparative view of housing choices in controlled and uncontrolled housing markets. *Urban Studies, 26,* 457–468.

Earnhart, D. (2002). Combining revealed and stated data to examine housing decisions using discrete choice analysis. *Journal of Urban Economics, 51,* 143–169.

Engel, J. F., Blackwell, R. D., & Miniard, P. W. (1995). *Consumer behavior.* Orlando, FL: Dryden Press.

Fawcett, J. T. (1986). Migration pyschology: New behavioral models. *Population and Environment, 8*(1&2), 5–14.

Fishbein, M., & Ajzen, I. (1975). *Belief, attitude, intention, and behavior: An introduction to theory and research.* Reading, MA: Addison-Wesley.

Fokkema, C. M. (1996). *Residential moving behaviour of the elderly: An explanatory analysis for the Netherlands*. Amsterdam: Thesis Publishers.

Fuller, T. D., Lightfoot, P., & Kamnuansilpa, P. (1986). Mobility plans and mobility behavior: Convergences and divergences in Thailand. *Population and Environment, 8*(1/2), 15–40.

Galster, G. (1996). William Grigsby and the analysis of housing sub-markets and filtering. *Urban Studies, 33*, 1797–1805.

Gibler, K., & Nelson, S. (2003). Consumer behavior applications to real estate education. *Journal of Real Estate Practice and Education, 6*(1), 63–83.

Goetgeluk, R. (1997). Bomen over Wonen: Woningmarktonderzoek met beslissingsbomen (Dissertation, the Netherlands Geographical Studies, Utrecht), p. 235.

Goetgeluk, R. W., Hooimeijer, P. (1991). *Household formation and success on the housing market in the Netherlands*. Paper prepared for the conference on Housing Policy as a Strategy for Change, Oslo, Norway

Goetgeluk, R. W., Hooimeijer, P., Dieleman, F. M. (1992). *The effectiveness of housing search: The role of motives for moving and housing market adjustment*. Conference on European cities: Growth and decline, Den Haag, The Netherlands.

Gregory, R., Lichtenstein, S., & Slovic, P. (1993). Valuing environmental resources: A constructive approach. *Journal of Risk and Uncertainty, 7*, 177–197.

Holland, J. H., Holyoak, K. J., Nisbett, R. E., & Thagard, P. R. (1986). *Induction. Processes of inference, learning and discovery*. Cambridge: MIT Press.

Hooimeijer, P. (1994). Hoe meet je woonwensen? Methodologische haken en ogen. In I. Smid & H. Priemus (Eds.), *Bewonerspreferenties: Richtsnoer voor investeringen in nieuwbouw en de woningvoorraad* (pp. 3–12). Delft: Delftse Universitaire Pers.

Hooimeijer, P. (2007). *Dynamiek in de derde leeftijd, de consequenties voor het woonbeleid*. Den Haag: VROM Publicatie.

Huber, O. (1986). Decision making as a problem solving process. In B. Brehmer, H. Jungermann, P. Lourens, & G. Sevon (Eds.), *New directions in research on decision making* (pp. 109–138). Amsterdam: North-Holland.

Huber, O. (1989). Information-processing operators in decision making. In H. Montgomery & O. Svenson (Eds.), *Process and structure in human decision making* (pp. 3–21). New York: Wiley.

Janis, I. L., & Mann, L. (1977). *Decision making. A psychological analysis of conflict, choice and commitment*. New York: The Free Press.

Kok, J. (2007). Principles and prospects of the life course paradigm. *Annales de Démographie Historique, 1*, 203–230.

Lindberg, E., Gärling, T., & Montgomery, H. (1988). People's beliefs and values as determinants of housing preferences and simulated choices. *Scandinavian Housing and Planning Research, 5*, 181–197.

Maclennan, D. (1977). Information, space and measurement of housing preferences and demand. *Scottish Journal of Political Economy, 24*, 97–115.

Manski, C. (1999). Analysis of choice expectations in incomplete scenarios. *Journal of Risk and Uncertainty, 19*(1–3), 49–65.

Mulder, C. H. (1993). *Migration dynamics: A life course approach*. Amsterdam: Thesis Publishers.

Mulder, C. H. (1996). Housing choice: Assumptions and approaches. *Journal of Housing and the Built Environment, 11*, 209–232.

Mulder, C. H., & Hooimeijer, P. (1995). Moving into owner-occupation. *Journal of Housing and the Built Environment, 10*, 5–25.

Musterd, S. (Ed.). (1989). *Methoden voor woning- en woonmilieubehoefte onderzoek*. Amsterdam: SISWO.

Newell, A., & Simon, H. A. (1972). *Human problem solving*. Englewood Cliffs, NJ: Prentice-Hall.

Payne, J. W., Bettman, J. R., & Johnson, E. J. (1992). Behavioral Decision Research - a constructive processing perspective. *Annual Review of Psychology, 43*, 87–131.

Payne, J. W., Bettman, J. R., & Johnson, E. J. (1993). *The adaptive decision maker*. Cambridge: Cambridge University Press.
Priemus, H. (1984). *Nederlandse woontheorieen. Volkshuisvesting in theorie en praktijk*. Delft: Delftse Universitaire Pers.
Ritchey, P. N. (1976). Explanations of Migration. *Annual Review of Sociology, 2*, 363–404.
Rossi, P. H. (1955). *Why families move: A study in the social psychology of urban residential mobility*. Glencoe: Free Press.
Sabagh, G., Vanarsdol, M. D., & Butler, E. W. (1969). Some deteriminants of intrametropolitan residential mobility – conceptual considerations. *Social Forces, 48*(1), 88–98.
Savage, L. J. (1954). *The Foundations of Statistics*. New York: John Wiley.
Simon, H. (1991). Bounded rationality and organizational learning. *Organization Science, 2*(1), 125–134.
Simon, H. A. (1977). *The new science of management decision* (Rev. ed.). Englewood Cliffs, NJ: Prentice-Hall.
Simon, H. A., Dantzig, G. B., Hogarth, R., Plott, R. C., Raiffa, H., Schelling, T. C., Shepsle, K. A., Thaler, R., Tversky, A., & Winter, S. (1987). Decision making and problem solving. *Interfaces, 17*, 11–31.
Slovic, P. (1995). The Construction of preference. *American Psychologist, 50*, 364–371.
Smid, I., & Priemus, H. (1994). *Bewonerspreferenties: Richtsnoer voor investeringen in nieuwbouw en de woningvoorraad*. Delft: Delftse Universitaire Pers.
Svenson, O. (1979). Process descriptions of decision making. *Organizational Behavior and Human Performance, 23*, 86–112.
Timmermans, H., Molin, E. J. E., & van Noortwijk, L. (1994). Housing choice processes: Stated versus revealed modelling approaches. *Journal of Housing and the Built Environment, 9*(3), 215–227.
von Neumann, J., & Morgenstern, O. (1947). *Theory of games and economic behavior* (2nd ed.). Princeton, NJ: Princeton University Press.
Vriens, M. (1997). Het meten van preferentie Structuren. In *Jaarboek van de Nederlandse Vereniging voor Marktonderzoek en Informatiemanagement* (pp. 241–258). Haarlem: De Vrieseborch.

Chapter 2
Traditional Housing Demand Research

Harry J.F.M. Boumeester

2.1 Introduction

Moving house is never a goal in itself, rather a means to achieving a goal (Goetgeluk 1997). In modern society there are general life goals towards which everyone strives, such as physical well-being and social acceptance (Lindenberg 1990). People tend to focus on various specific, more concrete goals, with the ultimate aim of achieving these general life goals (Coleman 1990; Oskamp 1997). Mulder (1993) refers in this context to preferences, which differ from person to person and can also change over the period of an individual's life (Willekens 1989; Oskamp 1997).

Moving house can be regarded as an adjustment in response to the altered housing needs of a household. Information about the current and future composition of the population in terms of households and the desired housing situations (housing preferences) provides a good foundation for mapping out current and future housing needs. Such knowledge can also be used by policy-makers and builders as a starting point in their acting.

Since the Second World War, the production of housing in the Netherlands has been heavily guided by central government. It is therefore important that the government has a clear picture of the current and future demand for housing. The Netherlands has a long tradition in researching housing needs – since 1977 the national Housing Demand Survey (*Woning Behoefte Onderzoek*: WBO) has taken place every 4 years. In 2006, this was changed to every 3 years and related subjects were included in the research (Housing Research Netherlands survey, *Woon Onderzoek Nederland*: WoON). The findings from this housing demand research is representative for the whole of the Netherlands and for all segments of the housing market. Since 1995, there has also been, initially annually and since 2000 bi-annually, national research

H.J.F.M. Boumeester (✉)
OTB Research Institute for the Built Environment, Delft University of Technology, Delft, The Netherlands
e-mail: h.j.f.m.boumeester@tudelft.nl

S.J.T. Jansen et al. (eds.), *The Measurement and Analysis of Housing Preference and Choice*, DOI 10.1007/978-90-481-8894-9_2,

carried out into the housing preferences of households with an above-average income. This research is commissioned by the *Nederlandse Vereniging van voor ontwikkelaars en bouwondernemers* (NVB) – the association for developers and builders.

Surveys into housing also take place in England – the national Survey of English Housing (SEH) (renamed the English Housing Survey in 2008) has been carried out annually since 1993. In the USA, the American Housing Survey (AHS) began in 1973, initially annually, and since 1983 has been carried out bi-annually.

The housing demand survey gives a picture – or "photo"– of the current housing distribution and expected housing demand in the short term at a given time. There is no indication, however, of how specific the picture is. By carrying out the research in individual regions, a "photo-mosaic" can be created by putting the individual pictures together. The geographical reference gives the findings more value. However, such a mosaic still only represents the housing market at a given moment and cannot be used to determine whether possible developments in the housing demand have taken place. For that, a succession of mosaics is necessary – transforming a "photo-mosaic" into a "film" and thereby giving the findings a reference point in time. There are two methods available that provide successive measurements of housing demand. On the one hand, there is the longitudinal approach, where a specific group (panel) of households is followed over time. The film is actually a serial, like a TV series with the same cast. De Groot discusses this approach in more detail in the chapter on longitudinal analysis. On the other hand, the same research is carried out at different moments in time with different, but representative, groups. An example of this approach is provided later on in this chapter.

In Sect. 2.2, research into housing preferences is brief looked at, as this provides the foundation for accurately determining the demand for housing by housing consumers. In Sect. 2.3, the procedure for the move from measuring housing preferences to housing demand research is briefly explained, and the WBO research is used to illustrate and clarify this process. Section 2.4 provides a state of the art of studies that have applied the transverse, or cross-sectional, approach in the analysis of housing demand. This includes both national and international studies creating a "photo" or "photo-mosaic." Following on from that the transition is made in Sect. 2.5 to a "film" by providing an example using data from repeated cross-sectional studies into housing demand in the Netherlands. Data are used from the demand for owner-occupied houses at the top end of the housing market in the period 1975–1997.

2.2 Housing Preferences Research

People's acting and thinking are often based on a long-term vision in order to provide continuity and security in life. Current behavior is adapted according to a person's long-term preferences. "Living is the continuous attempt to find concord between the current housing situation and one's aspirations, determined by considering the available practical possibilities, and the continuous attempt to find concord between one's aspirations and one's subjective ideal, determined by the

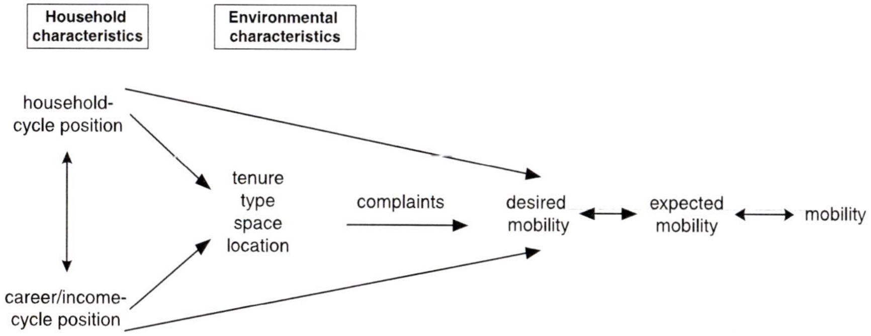

Fig. 2.1 The modified career/life cycle model (Source: Priemus 1984, p. 113)

available theoretical possibilities" (*Priemus* 1969, p.14, translation author). The individual endeavors to give his/her life shape according to fairly consistent paths, denoted as careers. People can furthermore follow parallel, strongly connected careers for different areas of their lives, such as education, work, sport, and creating a household and living (Mulder 1993; Mulder and Hooimeijer 1995; Clark and Dieleman 1996).

Every household has a specific housing need, which is partly determined by the status of the other careers. Changes in the household cycle or work cycle lead to changes in the housing needs. If the current housing situation deviates too much from the altered needs then this can lead to dissatisfaction. Once a certain threshold value has been crossed, people begin to think about moving to a dwelling that better suits their needs (Priemus 1984). Once the right kind of dwelling has been found and is available, a move to the more fitting dwelling will take place. The conceptual framework of housing preference research is based on these suppositions and is illustrated in Fig. 2.1.

2.2.1 *Preferences and Value: Compositional Modeling*

If people strive towards a particular goal they will want a dwelling and living environment that suits that goal. It is assumed that the dwelling and the living environment are made up of a collection of features (attributes). Different people with diverse goals will therefore ascribe different values to these attributes. People have a preference for those attributes to which they ascribe greater value; in other words, people have preferences. Working out the preference structure is central to the research.

According to some, the preference structure can be researched by studying the choices that people make. This assumes that the choice is a good reflection of a person's preferences. This type of research is known as "revealed preference" research. A number of important assumptions have to be made in order for the choice to be a valid source for measuring preferences. In a market where the consumer is boss and the suppliers react to the demand, the choice that is made is the perfect basis upon which to determine the preference structure. However, in a tight market where people probably have to make do with something that does not fit their criteria so well, the actual choice is not a good reflection of the preference. In that case, studying the preferences using a "stated preferences" approach is probably better. The problem here is then that the predictive ability of the preferences may be less realistic. The chapters by Koopman, De Groot and Goetgeluk, as well as the concluding chapter, provide greater insight into the question of stated and revealed preferences.

2.2.2 *Variables*

The stated housing preference is the combination of all the desired features of the dwelling as well as the location (the living environment). These various features are each given a separate value by the household when they are looking for somewhere to live. All these values together form a "total value" for the situation. The number of features is in theory endless; in practice (both in the actual search and in housing preference research), however, the number of features that individuals take into consideration is restricted. Housing preference research focuses on the "part-value" of each individual feature, whereby the total value of a dwelling with certain features can be determined. In such research the features mentioned in Table 2.1 are often utilized (the list is not exhaustive).

The total value can be determined by summing all the part-values and is an indicator of the desirability of the dwelling by the participant consumer. An idea of

Table 2.1 The most often utilized dwelling and environment features in housing preference research

Dwelling features	Environment features
Type of dwelling	Type and size of local council
Number of rooms	Type of neighborhood
Size of living room	Type of housing
Total usable surface area of dwelling	Period built
Backyard present?	Amenities in the neighborhood
Size of backyard	Public transport
Presence of balcony	Green and water
Size of balcony	Semipublic area (parking, etc.)
Storage space	Parking places
Tenure	Safety, including traffic
Price	Space, building density
Architecture	Urban development design for the neighborhood
Quality/level of maintenance	
Year built/period built	
Private parking place	

Source: Goetgeluk (1997); Heins (2002); Boumeester et al. (2005)

Table 2.2 Frequently used features for households in housing preference research

Demographic features	Social-economic features
Type of household (single, couple, family, single-parent family)	Position in the job market (self-employed, employed, unemployed, unable to work, retired early, retired)
Number of people	Number of hours work per week
Ages of household members	One or two incomes
	Level of income head of the household/partner
	Components of income for head of the household/partner
	Receipt of benefits/allowance
	Level of education
	Political stance

Source: Clark et al. (1990); Mulder (1993); Boumeester (2004); De Groot et al. (2008)

the "popularity" of a particular dwelling can be obtained by considering all the total values together. However, the popularity of a dwelling appears to vary between households with different dwelling needs and positions in the job market. The same dwelling can have a completely different value for one household than for another. A dwelling can also have different total values for the same household over time, if the household itself enters a different phase or the job market position changes. Housing preference research also needs to gather information about the demographic and social-economic features of households. Table 2.2 gives an overview of features frequently used (the list is not exhaustive).

The housing preferences are therefore closely linked to the features of the household. A family with children, for example, is more likely to prefer a single-family dwelling with a backyard than a single person is. Alternatively, a household with a high income is more likely to buy a dwelling than a household with a low income is. However, knowledge about the popularity of a dwelling does not tell us everything about possible plans to move. A household that tries to achieve maximum value will only make plans to move when it is clear that the current housing situation is valued less than other possibilities. In order to be able to utilize the findings about housing preferences in, for example, the calculation of the housing demand, insight into possible discrepancies between the current and the desired housing situation are therefore also important. The dwelling and environment features of the current housing situation therefore also need to be determined.

2.2.3 Data Collection

A household's "stated" housing preferences only become meaningful when they can be compared with the current housing situation (dwelling and environment). This is also true for the "revealed" housing preferences in relation to the previous housing situation. In addition, the preferences need to be studied in relation to the demographic features (type of household, number and age of people) and the social-economic features (income, education, type of income, position in the job market, number of earners) of the household.

In the WBO/WoON research, therefore, a questionnaire is used that comprises the following sections:

1. Extent of likelihood to move house/recently moved.
2. Features of the present dwelling.
3. Features of the present living environment.
4. Desired features of the dwelling.
5. Desired features of the living environment.
6. Features of the previous dwelling (for those who have recently moved).
7. Features of the previous living environment (for those who have recently moved).
8. (Current and expected) demographic features of the household.
9. (Current and expected) social-economic features of the household.

The form and structure of the utilized questionnaire in much housing preference research will be very similar to this structure used in the WBO. The WBO questionnaire comprises a number of closed questions, many with predetermined answer categories.

The WBO questionnaire is carried out using face-to-face and telephone interviews with 60,000 individuals who are representative of the Dutch population aged 18 years and over who are not living in institutions. The criteria for the participant group was altered and improved during the transition to the WoON in 2006 and the number of respondents was restricted. The two methods of interviewing give the interviewer the possibility to provide clarifying information in the course of the interview if necessary. Furthermore, due to the personal approach, the level of response and the distribution of the net response over the a priori formulated strata can be better monitored and more quickly adjusted. The methods used clearly differ in this respect to using written questionnaires, which can be offered either by post (hard copy) or via the Internet (digital copy). The main advantage of using written questionnaires is the financial saving in comparison with personal or telephone interviews (Hilkhuysen 1999).

It is clear that standard housing preference research is characterized by quantitative data collection. The features to be studied and possible answer categories (attributes and attribute categories) contained in the questionnaires are selected on the basis of the available knowledge. The influence of the researcher is restricted to the creation of the structured questionnaire. A large number of respondents are approached to take part, in order to obtain enough data to test the expected links between the features of the household, the dwelling, and the environment, frequently with a statistical analysis.

2.3 From Housing Preferences to Housing Need Research

Measuring housing preferences provides insight into how diverse households would like to live. Such insight provides, in combination with knowledge about expected changes in demographic and social-economic factors, information about the existing housing demand. Knowledge about the size and composition of this housing demand provides a good foundation for different decision-making processes. For example, insight into the quantitative (size) and qualitative (composition) housing need is necessary in order to be able to make well-founded decisions when drafts are made for the interpretation of the urban development plans of a location for new houses, or redevelopment areas, or for realizing the vision for a particular area at a particular time.

Housing demand research is therefore carried out in order to obtain information about the housing market in the short term (for example, mobility effects resulting from new-build) and in the long term (for example, whether there are structural changes in the houses that are sought after) and possibly to guide the market (government or market party). Housing demand research therefore needs to measure the correct housing preferences of all potential and actual households in the particular (local, regional, national) housing market.

2.3.1 Reliability and Validity

The measured preferences need to be reliable and valid. Reliability is based on accuracy. A questionnaire must produce the same findings when it is done the second time. Errors (for example, due to respondents guessing an answer) must be random rather than systematic. Guesses as answers are not a problem, as long as the guesses are not systematically too high or too low (systematic errors).

Validity is concerned with whether we are actually measuring what we thought was being measured. Are the measurements valid for the "concept as it was intended?" Is the variable properly operationalised and the correct question utilized? In other words, are the findings valid?

2.3.2 Selecting the Sample

It is clearly not feasible to actually measure the housing preferences of all households (that is, the whole population). The housing demand is therefore based on the opinions of a representative group of households from the population. A random sample is drawn from the total group of households and their housing needs are ascertained. The sample is made up of all the individuals or households for whom the necessary information is available and the group of respondents is drawn from this sample. A random sample is when each household in the sample group has an equal chance of being chosen. The straightforward random sample and the stratified sample are the most utilized.

The larger the diversity in the features of the individuals, the larger the sample from the group needs to be to obtain reliable findings. By dividing the group into several more homogeneous groups (strata), such as age or income groups, or position in the job market, the diversity within each stratum is reduced. A simple random sample can then be drawn from each stratum, which then has a smaller size. This is known as a stratified sample.

In the WBO a sample is created of all individuals who are 18 years or older in the Netherlands and who are registered with their local council (in the *Gemeentelijke Basis Administratie* – GBA). From this group a stratified sample can be drawn according to the design of the survey, divided according to age, marital status, country of birth, and size of council. In the WBO from 2002 there were in total 24 strata, or clusters (VROM 2003).

2.3.3 Size of Sample and Nonresponse Percentages in Housing Research

The size of the sample is partly dependent on the desired reliability of the sample findings and the expected diversity within the wider sample. The size of the sample as well as the nonresponse level determine the expected net response, in other words the number of individuals who actually complete the questionnaire in the research. Not everyone will be prepared or will want to take part in the research. The researcher needs to take, as far as possible, the nonresponse into account when determining the size of the sample. The nonresponse is often estimated based either on experience from previous research or by doing a small pilot study. The nonresponse percentage varies greatly depending on the target group (for example, young people or old, looking for new accommodation or not likely to move), the subject (very general or more specific) and the manner of the interview.

In the WBO in 2002, the regular sample comprised approximately 100,000 individuals: 61% took part in either a face-to-face or telephone interview, 24% refused, and the other 15% could not be contacted. Individual councils had the option to resample if desirable. Interviewing extra individuals made it possible to perform reliable analyses at the council level. A sample of at least 50,000 individuals was drawn for this resampling, 59% of whom were interviewed. This national study ultimately involved 92,000 respondents (VROM 2003).

2.3.4 Generalizability of the Findings: Weighting

Even if stratified samples are utilized, it is not always possible to generalize the findings of the sample to the general population. If the different strata are not proportional then, just as with straightforward sampling, the findings need to be "weighted." This can also happen when the a priori stratification goals are not reached (for example, due to deviant response percentages that were insufficiently adjusted). This latter procedure is, however, risky as the composition of the nonresponse is often not known. Each case in a stratum is given a weight such that the proportion in numbers between the strata in the total response group corresponds with the original proportion in the empirical population (the original sample group).

Weighting of the findings is not to be confused with adjusting the results. By adjusting is meant that the weighted number of cases in the response group is multiplied, such that this number corresponds with the total number in the population (this does not change the division into features). This is necessary in order to express, for example, the total housing demand (or the balance of housing demand and stock of dwellings) into actual number of dwellings. The 92,000 respondents in the WBO from 2002 are representative of the 16.1 million inhabitants, 6.9 million households, and 6.6 million dwellings in the Netherlands (VROM 2003).

2.4 Cross-sectional Analysis

The findings from housing preferences or housing demand research are frequently used to map out the current housing market, on the basis of which local, regional, or national policy can be developed for different areas (distribution of land for housing, restructuring, housing expenditure, supply of new-build). Frequently it is sufficient to do descriptive analyses such as frequency distributions, two or more dimensional contingency tables and comparisons of subsets using center, distribution, and skewness measures. The findings are presented in tables or graphically.

The descriptive analyses can be carried out for a number of subjects, based on the research criteria. As such, a more specific description can be drawn up of the composition of the population from the demographic and/or social-economic features of the households. Types of households can also be drawn up by combining a few features. Equally, the existing housing stock can be sketched in more detail by making use of the possibility of combining diverse features from the dwelling and living environment. In this way different housing products can be distinguished, often using features such as tenure, type of dwelling, size, price and/or location (Moore, and Clark 1990; Kruythoff 1993; VROM 2002; Boumeester et al. 2004).

An important added value from the housing demand research lies in the fact that the actual division of living space can be seen in these results. It is otherwise not possible to make such a direct link between household features and dwelling features from the available population registration and housing statistics.

By dividing the households and dwelling supply into types, it is possible to distinguish groups of housing consumers: groups of more or less comparable households with similar housing situations (Clark and Dieleman 1996; Oskamp 1997; Boumeester 2004; VROM 2007a).

For households who have moved house in the last 4 years, the previous housing situation is also mapped out. Based on the information about both the previous and the current dwelling recent relocation movements can be more closely analyzed. The size of the flows can be centrally determined, and also the nature of the relocation. This involves splitting the different flows into housing market areas (regions or housing areas), housing market segments, or housing consumer groups. By linking the information from housing consumer groups who have recently moved house with the features of the dwellings involved, housing demand profiles can be constructed (Boumeester 2004; Kulberg and Ras 2004; VROM 2007b; Boumeester et al. 2009). The flow effects can be determined by linking the information about the dwelling concerned to the features of the previous dwelling and thereby also the relocation chains. With this information, the expected dynamics of the housing market as a result of new-build can be modeled (Everaers 1990; Teule 1996; Goetgeluk 1997; Elsinga and Goetgeluk 2003; De Groot et al. 2008).

Respondents in the housing demand research are also asked if they are likely to move house: whether they have more or less concrete plans to move in the near future (1–2 years). If the likelihood that a household will move is high, further questions are posed about the features of the desired housing situation. This information can be used to map out the popularity of particular housing products and can possibly be split into diverse housing consumer groups. Potential mobility flows can be analyzed by linking the desires of a respondent with the features of their current housing situation. Furthermore, once again the desired mobility flows can be distinguished into housing market areas (regions, districts, or living environments), housing market segments (rent or owner-occupancy, type of housing, price group), and housing consumer groups (desire profile). Potential mobility balances can often be drawn up, or the theoretical balance calculated, whereby insight can be gained into the potential demand in specific housing market segments (Elsinga and Goetgeluk 2003; Boumeester and Van der Heijden 2004; VROM 2007c; Boumeester et al. 2008; De Groot et al. 2008). Figure 2.2 illustrates an example of a mobility balance and a theoretical shortage of dwellings.

An indication of the future housing demand can be determined by applying the desired (or current) distribution of housing space in a research area onto the projected number of households for the same area. This can only be indicative, as such an approach involves making a number of serious assumptions: that the households remain or come and live in the area, the preferences of diverse types of households remain unchanged over time, the relocation movements of households is carried out in agreement with each other and there is no substitution of housing preferences.

2.4.1 Repeated Cross-sectional Analysis

As stated earlier, a one-off survey of the housing preferences within a housing market area (national, regional, or local) only gives insight into the situation in that particular area. The data collected in a one-off study into housing demand can be used to more accurately specify the housing supply (market segments) and the housing demand (consumer groups), as well as to determine both the current and desired housing space distribution and the discrepancy between them. However, there is no reference point for the results of the analysis.

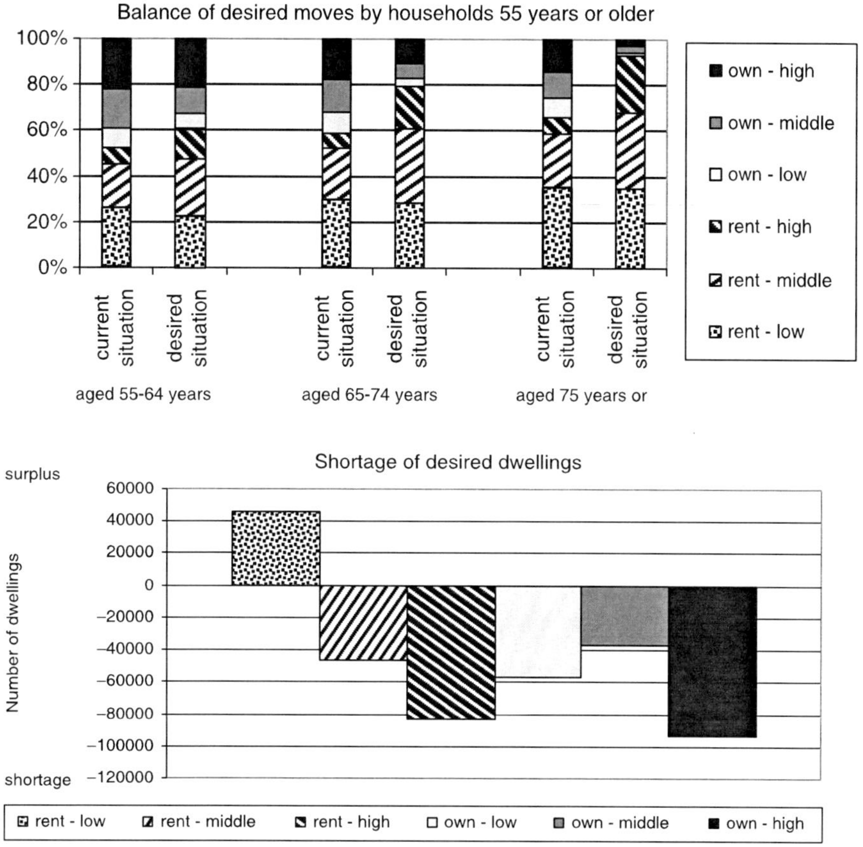

Source: WBO 2002 (OTB/TU Delft calculations)

Fig. 2.2 Examples of the balance of desired moves and the potential shortage of desired dwellings in 2002 (Source: WBO 2002 (OTB/TU Delft calculations))

For this reason, the survey used is often repeated in other housing market areas (or in a national study which is split up into different regions) or at another time in the same housing market area. Utilizing the same measurement instrument (the same questionnaire) enables a direct comparison of the findings and the relative value of the results can be determined.

Cross-sectional data become even more valuable if the survey is repeated at regular intervals in the same research area with the same measurement instrument. In this way a succession of findings regarding unequivocal, operationalised variables are created at an aggregated level. It is then possible to determine changes over time for particular variables, as well as the links between variables. Possible trends can then be identified. Such a survey in the Netherlands is the national Housing Demand Survey (1977, 1981, 1985/6, 1989/90, 1993/4, 1997/8 and 2002) and its successor the Housing Research Netherlands Survey 2006.

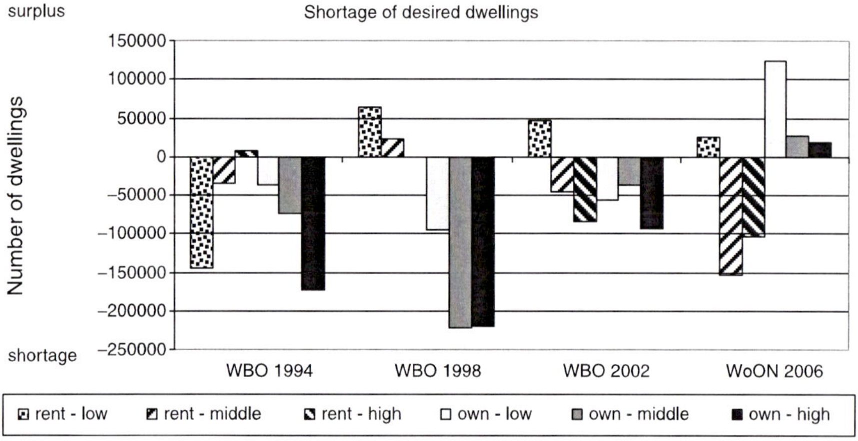

Fig. 2.3 The potential shortage of desired dwellings in 1994, 1998, 2002, and 2006 (Source: WBO 1993/94, 1997/1998, 2002, and WoON 2006 (OTB/TUDelft calculations))

In addition to this, since 1995 there has been the recurring national survey Housebuyers in Profile (HiP 1995, 1996, 1997, 1998, 1999, 2000, 2002, 2004, 2006 and 2008) that specifically studies the housing situation and wishes of households in the Netherlands with an above-average income. In the publications from this latter survey, much use is made of the desire profiles (Boumeester et al. 2008).

It is possible, with the help of the repeated cross-sectional data, to determine changes or trends in housing preferences and housing demand. For clarity, Fig. 2.3 illustrates once more the potential shortage in desired dwellings in 2002, this time in combination with the same information from surveys before that time and from 2006. It is clear that the potential housing shortage in 2002 would be interpreted very differently based on Fig. 2.3 than on Fig. 2.2. Looking only at the potential housing shortage in 2002 it is possible to conclude that there is great shortage of expensive houses for sale (100,000 dwellings). Figure 2.3 indicates, however, that this shortage is considerably smaller than in the 1990s, and can even be regarded as part of a transition towards a surplus in 2006.

The WBO research asks households that have recently moved (in the 4 years before the survey) in which 6-month period they last moved. The features of both the previous dwelling and the current dwelling for this last move are determined. With the help of successive WBO data files, a variable can be constructed that contains the number of house moves for each 6-month interval over a long time period (first half of 1975 to second half of 2009). Such variables are often used for secondary data analysis, including for research into the demand for more expensive houses in the Netherlands (Boumeester 2004). This research provides a good illustration of the techniques of repeated cross-sectional analysis, which we now move on to.

2.5 An Example of Repeated Cross-sectional Analysis: The Development of Demand in the Housing Market

Moving house can be seen as a decision-making process, where the current housing situation, the possibly changing housing preferences and the knowledge about other dwelling spaces play a role. The decision to move to a particular dwelling can be explained as comprising three part-decisions: the moment of deciding to move house, the choice between renting and buying and the choice for a certain amount of dwelling features (type, size, and quality of the dwelling). These part-decisions are closely interconnected. The decisions are often made all at the same time, and the choice in one part-decision often determines the choices made for the other part-decisions (Laakso and Loikkanen 1992; Elsinga 1995). The choice for a particular sort of dwelling can lead, for example, to an individual entering the house-buying sector. If the desired dwelling is not available then substitution behavior may take place, where a move is made to a comparable rental dwelling or another dwelling is bought with features that are different to the desired ones. A decision might also be made either not to move house at all or to postpone moving.

The revealed housing choices of a household reflect the housing preferences, but also the housing market conditions within which these choices are made. It was concluded above, in Sect. 2.2, that housing preferences are closely linked to the demographic and social-economic features of a household. The actual demand for dwellings in a particular market segment is also linked to the demographic (age and type of household) and social-economic (household income, level of education) composition of the population. However, the availability, accessibility, and affordability of the desired dwelling also play a role in the eventual size of the demand for the dwelling (Hooimeijer and Linde 1988; Goetgeluk et al. 1992; Teule 1996; Boumeester 2004). Government policy and the policy of mortgage credit institutions are also structural explanatory factors. These factors have influence particularly in the long term on the development of demand, with the exception of the social-economic factors, which can also be of influence in the short and middle–long term. Far-reaching changes in policy can also play a role in the short term.

In addition, there are three factors regarding the dynamics of the housing market itself that can play a role in the development of demand for more expensive houses in the short and middle–long term. This is the supply of new-build dwellings as well as dwellings in the existing stock, the sale price of the dwellings (the actual price as well as the percentage of change) and the rental prices.

The housing policy of the Dutch government seemed to go one step further than in the twentieth century with the publication of the paper "People, Wishes, Living" (VROM 2001). Key words were "quality improvement" and "freedom of choice." The focus is on the housing consumer's qualitative housing demand, both in the rental sector and particularly in the owner-occupied sector. Realizing the policy goals is even more dependent on the actual housing market conditions than in the previous period. Such a policy does not only demand an accurate and current insight into the supply and demand in the housing market, but greater and more accurate knowledge about expected developments in the supply and demand in the near future are also of great importance. It is also important to know which factors influence the demand for housing and in what way. These developments in policy provided the stimulus for a study by Boumeester (2004) into the development of the demand for houses at the top end of the market in the Netherlands. Boumeester (2004) aimed to answer the following two research questions:

- Which factors influenced changes in the level of the actual demand for expensive houses between 1975 and 1993?
- To what extent can the actual demand for this sort of house be explained by demographic, social-economic and/or cultural changes at a macro level and through changes in the state of the housing market?

2.5.1 The Technique in General

A search through both the Dutch and the international literature revealed that there appears to be no explanatory model available specifically for the demand for expensive owner-occupied housing. It is therefore not possible either to utilize or refute an existing model. A new model has therefore been constructed, making use of the data available from different sources.

Several choices need to be made before a model can be constructed. Firstly, a choice regarding the content needs to be made to incorporate a "constant factor" in the model. Omitting a constant assumes that the dependent variable will have the value "zero," if there is one (in a multiplicative regression model) partial effect, or that all (in an additive model) partial effects equal "zero." This assumption is not made in the model for the demand for expensive houses and therefore a constant is incorporated into the model.

Given the exploratory character of the analysis that was carried out on the changes in demand referred to, it must equally be ascertained whether the effects of the independent variables on the dependent variables need to be delayed by one or more time periods. For example, an increase in interest rates at time "*t*" often leads to a change in the demand for housing at time "*t*+1" or "*t*+2." An increase in mortgage rates leads to a reduced loan capacity, whereby some of the potential house buyers are forced to start looking in another, cheaper, price category. These people will also choose and move to a house in that cheaper price category. This search and choice process takes time. The length of the delay can be determined by theoretical assumptions, but can also be determined by "trial and error" during the statistical estimation of the model. Just as in Boumeester (2004), a combination of both is often applied.

As stated earlier, constructing the model needs to be done as economically as possible (the greatest possible explanatory power with the smallest number of dependent variables). This requires that the model is constructed step by step, involving two possible approaches: either starting from a full (saturated) model and working backward to an optimal model, or starting with a model with only a constant and building up to the optimal model. In Boumeester's (2004) study the second approach is utilized. The analysis begins with an estimation of the values of the dependent variables using a model with only a constant. The partial correlations between the residuals are then determined from this estimation and all the possible explanatory variables. The first independent variable is added on the basis of the direction (meaningful) and strength (as high as possible) of the correlation coefficients. This extensive model is then used to estimate the dependent variables again. The partial correlation coefficients between the residuals and the remaining possible explanatory variables are calculated again, on the basis of which the next dependent variable can be selected and added to the model. This procedure is repeated as long as the addition of an independent variable leads to substantial improvement in the model. It is possible that the whole process of estimating the model can be repeated with alternative combinations of the explanatory variables in the model.

2.5.2 The Method

The aim of the analysis is to develop an explanatory model of how the demand for expensive owner-occupied housing evolves, simultaneously based on the changes in diverse social-economic, demographic, and housing market factors. This implies, therefore, a multivariate analysis that results in an analysis model very similar to a standard regression model (see, for example, Field 2004, p. 116):

$$Y = \beta_0 + \beta_1 X_1 + \beta_2 X_2 + \cdots + \beta_n X_n + \varepsilon_i \tag{2.1}$$

The same criteria hold for an assessment of this model as for a standard multiple regression analysis. Firstly, the model needs to be as sparing as possible; in other words a model with the smallest number of independent variables and the greatest explanatory power possible.

Both the explanatory variables incorporated in the model and the direction of the determined links need to be theoretically possible. If the model is also going to be used to gain insight into possible changes in the short or middle–long term then the "turning points" in the development of the demand need to be explained as well as possible.

From a statistical point of view, just as with a standard regression model, this model must meet the following criteria: the links between the dependent variables and the individual explanatory variables need to be linear (no *nonlinearity*), the independent variables must not be too strongly correlated with each other (no *multicollinearity*), the residuals must be normally distributed with an average of zero and the distribution must be the same for each value of the predictor(s) (no heteroskedasticity) and must not be correlated with each other (no *autocorrelation*) (Lewis-Beck 1980; Berry and Feldman 1985).

Due to the specific role of "time" in the model, the proportion of the explained variance, R^2, is assessed differently than in a standard regression model. R^2 indicates the correlation between the actually observed values and the estimated values of the dependent variable using the model. Due to the fact that the score in year (t) may also have an influence in subsequent year(s), known as the trend correlation, a high value for R^2 is generally pretty quickly obtained. In addition, R^2 is often estimated slightly too high in small sample sizes, giving the impression that the model fits better than is actually the case. It is therefore better to utilize R^2 corrected for the number of degrees of freedom. Just as in standard regression analysis, the analysis aims for an R^2 that is as high as possible (Draper and Smith 1981).

The standard error (standard error of the estimates, SEE) is aimed to be as low as possible. The smaller the SEE, the more accurate and reliable the model. The standard error is used to determine the 95% confidence interval of the estimated values (Draper and Smith 1981).

In repeated cross-sectional analysis, one has to take possible autocorrelation into account. Autocorrelation indicates a connection between observations at a particular time with observations from the same series at an earlier time. If autocorrelation is reported, then consecutive observations are not independent of each other. Autocorrelation can lead to incorrect standard errors and consequently incorrect hypothesis tests and confidence intervals (Cohen et al. 2003). To test for autocorrelation the Durbin–Watson test (DW) is often used. The value of the DW statistic *d* always lies between 0 and 4 (Maddala and Lahiri 2009). Values close to zero indicate a positive autocorrelation between the residuals and values near 4, a negative autocorrelation. A value of 2 indicates no autocorrelation. A more precise indication can be obtained by comparing the value *d* to lower and upper critical values, which are based on the number of parameters (predictors + constant) and the number of cases in the model (see, for example, Table A.5 on p. 616 of Maddala and Lahiri (2009)).

A last comment concerns the number of observations required for a reliable analysis. Although there is no standard rule, the number of observations in the example given – 38 observations (19 × 2 half-years) – seems to be somewhat limited. This is a limitation of the current example.

2.5.3 The Variables

The dependent variable "realized demand for more expensive owner-occupied houses" can be transformed into the number of households that choose a dwelling in the more expensive segment of the market, and therefore choose not to rent or to buy a cheaper house. In the literature, it is clear that changes in the sale price of dwellings can influence the buying behavior of housing consumers, in terms of both affordability and the so-called speculative effect. The variable "sale price" definitely plays a role in the explanation of the progression of the level of demand for more expensive houses to buy. If the average sale price is utilized to distinguish the more expensive segment, then part of the explanatory power will not be seen. For this reason, the term "more expensive houses to buy" is defined as "all detached and semidetached dwellings in the owner-occupied sector" (Boumeester 2004). Closer examination shows that the average sale price for these types of dwellings is higher than for other types of dwellings. Detached and semidetached dwellings can therefore be considered as a proxy for the top end of the owner-occupied sector. The housing market sector "detached and semidetached dwellings" includes at least 45% of the housing market (which is about 20% of the total housing stock).

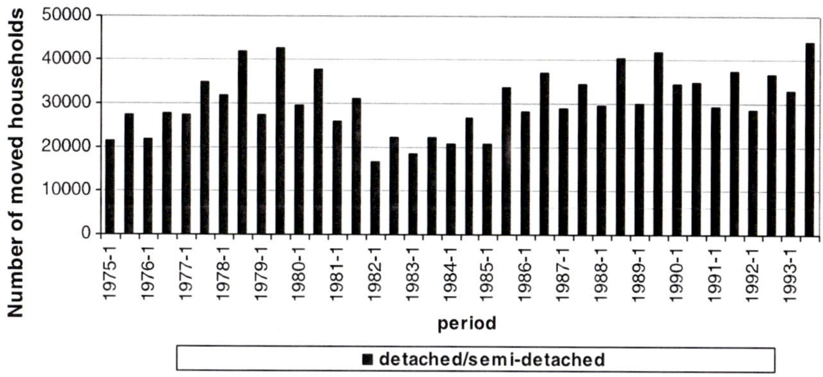

Fig. 2.4 The number of households that moved to a detached or semidetached dwelling in the Netherlands, on a half-year basis, during the period 1975–1993 (Adapted from Boumeester 2004, Fig. 8.1, p. 217)

Changes in the realized demand can be determined with the help of the consecutive WBO databases. Using these data, a series of the number of house moves on a 6-monthly basis can be constructed over a longer period of time (1975–1997). Figure 2.4 illustrates the changes in the demand for more expensive houses and this demand for detached or semidetached houses to buy.

When there is no insight into the changes in the total number of households, the developments within this series do not say much about the changes in the development of demand in the highest segment of the house-buying market. The number of moves into detached or semidetached owner-occupied houses is therefore related to the total number of households in the Netherlands. The variable to be explained therefore becomes "the percentage of all households that move to a detached or semidetached house per half year." As a consequence, the percentages are advanced by 2 months. Figure 2.5 illustrates the course of the dependent variable. The share varies from 0.75% in the second half of the 1970s, to 0.35% at the beginning of the 1980s to approximately 0.60% in the first half of the 1990s.

As was indicated in the introduction to Sect. 2.5, the literature shows that the actual relocation behavior is dependent on the housing preferences of households and on the availability, accessibility, and affordability of dwellings. The number of relocations to dwellings at the top end of the market in a particular period is therefore partly dependent on the demographic and social-economic composition of the population, the supply of such dwellings (both in new-build and the existing housing stock) changes in prices in the housing market and within the rental sector, and the possibilities within the mortgage market (level of mortgage interest rates, types of mortgage).

The potential predictors are therefore, in the first instance, constructed from the possible explanatory variables listed below. Series of data on a half-yearly basis are gathered for these factors and then the most appropriate way to represent the data is

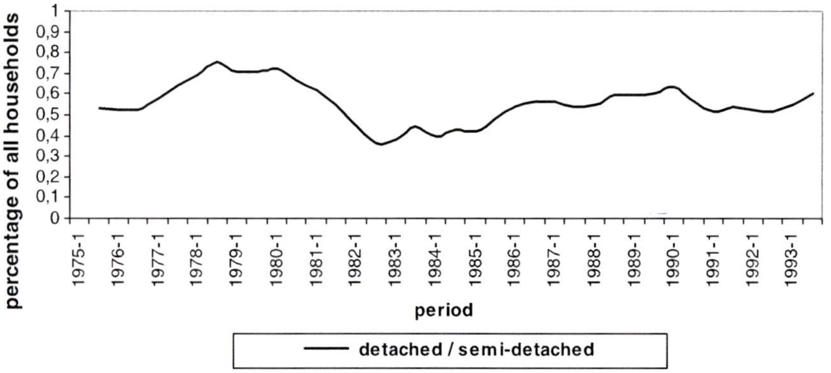

Fig. 2.5 The number of households that moved to a (semi)detached owner-occupied dwelling, as a percentage of the total number of households in the Netherlands, in the period 1975–1997 (two periods average) (Adapted from Boumeester 2004, Fig. 8.3, p. 224)

sought by applying transformations (absolute figures, percentage of change, advanced averages over two or more periods). The following datasets are involved:

- Population growth by age categories (absolute numbers).
- Average income of households (absolute value and percentage of change).
- Inflation.
- Average house price (nominal, real, absolute value, and percentage of change).
- Average rent for housing (nominal, real, absolute value, and percentage of change).
- Average mortgage interest rate (nominal, real, absolute value, and percentage of change).
- Newly built dwellings in the owner-occupied sector or in the rental sector (absolute numbers).
- Proportion of the owner-occupied sector in the existing housing stock (percentage).

By studying the correlations between the dependent variable and these factors, as well as by graphic comparisons, the usability of these factors in the time-series analysis can be determined. It is also taken into account that changes in these factors may have a delaying effect on how the demand changes. For example, from the literature it is already known that a fall of the mortgage interest rate or a bigger housing supply (newly build and in the housing stock) only affects the consumers' choices on the housing market after some time.

2.5.4 *Results*

After examining a number of alternative explanatory models, it seems that model A (see Table 2.3) can most accurately estimate the dependent variable "number of households that have moved into a detached or semidetached house, as a percentage

Table 2.3 Model of explanation for the volume of the demand for (semi)detached owner-occupied dwellings in the Netherlands in the period 1975–1993

Model profile[a]	b	t	B
Model A			
Constant	0.211	(3.60)	
Real price{1}	0.002	(10.92)	0.762
Interest rate{2}	–0.021	(–4.92)	–0.303
Supply (1}	0.00000489	(5.04)	0.260
Aged 65+	0.000011	(5.26)	0.310
Nominal price{3}	–0.004	(–3.30)	–0.232
N	37	Sign on 5%	
R^2	0.932		
SEE	0.025		
Durbin–Watson	1.850		

Source: WBO 1977/1978, 1981, 1985/1986, 1989/1990 and 1993/94 (OTB/TUDelft calculations)
b *b*-coefficient, *t* value of the *T*-test, *B* standardized *b*-coefficient, *real price* average real house prices (corrected for inflation, in terms of 1994 prices), *interest rate* average nominal mortgage interest rate, *supply* number of newly built dwellings in the middle and upper price classes, on a half-year basis (a two periods average), *Aged 65+* increase (absolute) of the number of persons in the age category 65 years or older, on a yearly basis (a two periods average), *nominal price* changes (%) in the average nominal house prices, on a half-year basis (a two periods average), *N* number of observations, R^2 percentage explained variance, corrected for the number of degrees of freedom and based on the number of observations and the number of independent variables in the model, *SEE* standard error of the estimated value

[a] The addition {1} means that the independent variable affects the dependent variable with a lag of one period (half year)

of all households." This involves a model that comprises five explanatory variables as well as the constant. Three variables are related to the housing market conditions, one variable is a social-economic indicator and the fifth variable is a demographic indicator. The negative coefficients indicate that the level of the mortgage interest rate and the variation from the average nominal sale price are negatively related to the size of the realized demand for more expensive housing. The datasets used can be found in Table A.1 of the Appendix.

The *t*-test makes clear that in this model all relations between the predictors and the dependent variable are statistically significant, meaning that the predictors have an influence on the dependent variable. The level of the average relative sale price has by far the strongest relation (standardized *B*-coefficient = 0.76) with the dependent variable to be explained. Increases in real house prices lead, other things being equal, to more moves to detached or semidetached houses, with a delay of one period (a half-year). Housing consumers also want to profit from the relatively large increases in the value of houses. However, there also appears to be a brake on these changes, given the negative correlation coefficient for the factor "change of nominal selling price" (*B*-coefficient = –0.232). If the actual price, corrected for short-term price fluctuations, continues to rise too quickly, then after a delay of three half-years there will be a small drop in demand. If the other explanatory factors stay the same, the affordability of the houses comes under pressure.

The level of the mortgage interest rate is included as the second variable in the explanatory model and in the final model as the third strongest relation (*B*-coefficient = −0.30) with the realized demand. The interest rate changes have a negative effect, with a delay of two half-years.

The relation between the changes in the supply of new-build houses and the actual demand appears to be less strong (*B*-coefficient = 0.26) than the mortgage interest rates, while at the same time the contribution to the total explained variance is slightly higher. The effect on the demand is seen with a delay of a half-year.

Finally, the demographic variable "increase in the number of individuals 65 years or older" appears to have a strong relation with the dependent variable (*B*-coefficient = 0.31). However, the relation does not correspond with the expected relation, given the positive correlation coefficient. There is apparently no demand effect, but rather a supply effect: more older people leads to more vacated detached and semidetached houses that other households can move into. In order to explain this we need to look at the great "shortage in the housing market" during the research period. The big potential demand from housing consumers for this type of housing can only be transformed into actual relocation when such houses become available, for example, due to older people moving house.

The model statistics indicate that the model is, on the whole, good. Approximately 93% of the variation in the dependent variable can be explained by this comparison. The standard error for the estimated comparison is low, indicating a good fit between the observed and estimated sets. In addition, the DW value (1.85) is sufficiently close to 2 to indicate that there is no autocorrelation. When comparing this value to the upper and lower critical levels, which are calculated on the basis of the number of parameters and the number of cases in the model, there is a statistical indication that the error terms are not positively autocorrelated. Note, however, that this statistic can be biased by the relatively small number of observations.

2.5.5 The Predictive Potential of the Explanatory Model

Based on the explanatory model presented for the actual demand for detached and semidetached housing it is possible to predict future changes in demand. An important condition is that the relations found in the period studied do not change in successive years. In addition it has to be assumed that the conditions for other factors, which are possibly of influence but not present in the model, remain unchanged.

In the study of Boumeester (2004) this was not the case, due to the introduction of the double-earners' mortgage in the Netherlands in 1992 that suddenly increased the number of financing possibilities for households Therefore an adjustment of the model was necessary by adding a dummy variable to distinguish the periods up to 1993 and after 1993. The predictive ability of that adjusted time-series model B, with only slightly changed *b*-coefficients for all other predictors, is determined for

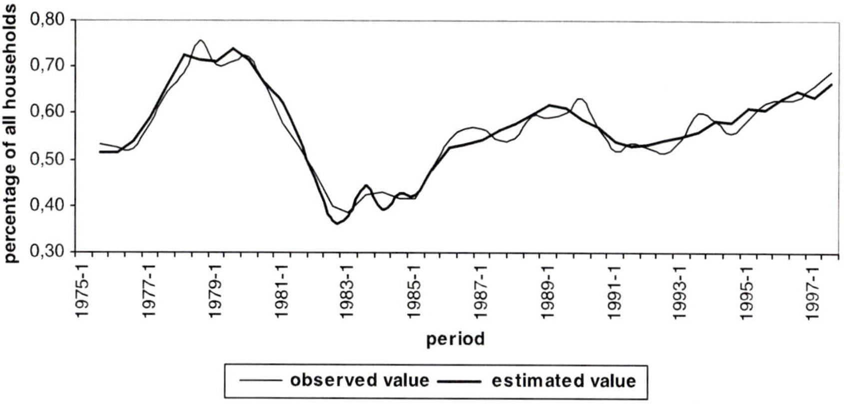

Fig. 2.6 Observed and estimated number of households that moved to a (semi)detached owner-occupied dwelling, as a percentage of the total number of households in the Netherlands in the period 1975–1997 (Source: WBO 1977/1978, 1981, 1985/1986, 1989/1990, 1993/94, 1997/1998 (OTB/TUDelft calculations))

the period 1975–1998. The result for the actual demand for detached and semidetached housing is represented graphically in Fig. 2.6 and the model data for the new model (model B) can be found in Table A.2 in the Appendix.

The explanatory model for the period 1975–1998, model B, can be written as:

$$D = 0.227 + 0.002 \times \mathrm{RP}_{t-1} - 0.023 \times \mathrm{MI}_{t-2} + 0.00000469 \times S_{t-1} + 0.00001 \times A$$
$$-0.004 \times \mathrm{NP}_{t-3} - 0.047 \times \mathrm{DUM}_{t-3}$$

where

D = number of movers to (semi)detached owner-occupied dwellings, as a percentage of the total number of households
RP = average real house prices (corrected for inflation, in terms of 1994 prices)
MI = average nominal mortgage interest rate
S = number of newly built dwellings in the middle and upper price classes, on a half-year basis (a two periods average)
A = increase (absolute) of the number of persons in the age category 65 years or older, on a yearly basis (a two periods average)
NP = changes (%) of the average nominal house prices, on a half-year basis (a two periods average)
DUM = period effect, with 1975–1992 = 0; and 1993–1997 = 1

The actual data for changes in the demand for detached and semidetached owner-occupied dwellings in the period 1994–1997 is available from the Housing Demand Surveys 1997/1998. Besides the changes in demand in the period up to 1998, the actual changes in the explanatory variables are also available. By adding these actual values to the regression equation in the explanatory model, the "predicted" change in demand for detached and semidetached owner-occupied dwellings can be determined for the named period. This prediction can then be compared with the actual change in demand, available from the consecutive Housing Demand Surveys.

As seen in Fig. 2.6, the actual and estimated courses of the demand are now more similar. The dummy variable is included in model B with a delay of three half-years and has a negative relation with the dependent variable. This dummy therefore corrects for the delayed reaction in the actual demand to the cited changes in the granting of mortgages. Comparing the *B*-coefficients in the original (see Table 2.3) and the new model (see Table A.2) shows that these are very similar, both in direction and in strength. This confirms the assumption that the partial relations between the independent variables and the actual demand are the same both before and after 1993.

2.5.6 Demand Projections Using Scenarios

Changes in the demand for detached and semidetached houses can be predicted for several years using the model, even when the actual values of the independent variables in the comparison are not available. An important assumption in this case is that the statistical relations are the same for these years as for the period 1975–1997. An estimation of the future values of the explanatory variables also needs to be made. In this situation scenarios are often utilized, within which the expected changes in the variables should show a logical, realistic relationship.

Working with scenarios usually aims to provide a broad range of the possible future changes in the dependent variable rather than providing as accurate a prediction as possible. The detailed scenarios need to be clearly distinguished from each other: a minimum and a maximum variant (and possibly a variant in between), a favorable and an unfavorable variant, or a most probable variant and a more desirable variant.

The expected demand for detached and semidetached dwellings in a scenario can be calculated by adding the values, per period, for the six independent variables from that scenario into the regression equation for model B. This calculation is repeated for each scenario and the findings can then be compared. A graphic representation is also often useful.

The quality of such predictions of housing demand therefore depends on the demonstrated validity/stability of the model, the degree to which the context stays the same and the credibility of the scenarios utilized.

Boumeester's (2004) study calculated a model-based prediction for the changes in the demand for detached and semidetached owner-occupied dwellings. Three scenarios were used: a very favorable future scenario (scenario C), an unfavorable future scenario (scenario B) and moderate development (scenario A). For the period 1998–2002, the actual values for the independent variables are utilized and from the first half of 2003 the model input is scenario-specific. An overview of the utilized data can be found in Table A.3 in the Appendix.

The results are illustrated in Fig. 2.7. The predictions according to the three scenarios clearly vary in the last years. The value of the dependent variable could

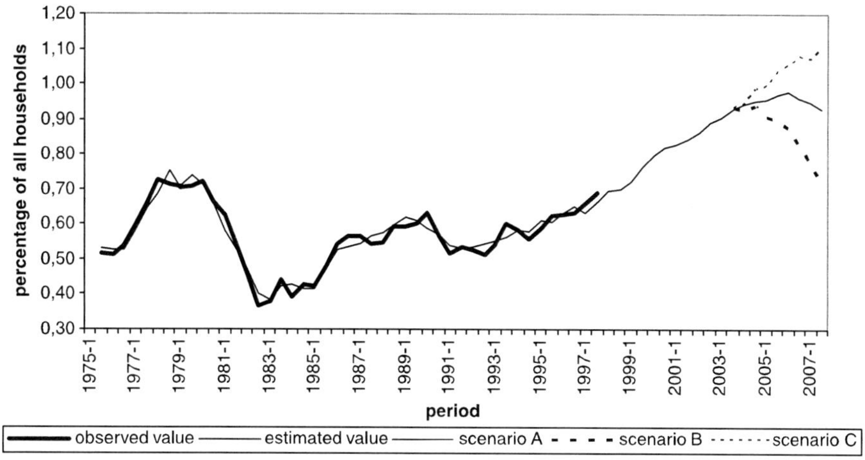

Fig. 2.7 Observed and estimated number of households that moved to a (semi)detached owner-occupied dwelling, as a percentage of the total number of households in the Netherlands in the period 1975–1997, and the forecast using three scenarios for the period 1998–2007 (Source: WBO 1977/1978, 1981, 1985/1986, 1989/1990, 1993/94, 1997/1998 (OTB/TUDelft calculations))

increase to 1.1% of all households in 2007 under the most favorable conditions, while in the most unfavorable scenario this value decreases to 0.72%. Converted into the number of households that would move into a detached or semidetached owner-occupied dwelling in the second half of 2007, the range lies between 81,000 and 52,000 households. On a yearly basis, this implies an actual demand for houses in this market segment in the region of 100,000–160,000 dwellings.

2.6 Conclusions

In this chapter, we have shown that the value of housing demand research lies in the connection between the features of households on the one hand and the features of the current housing situation (the actual division of housing space) and the housing preferences of the household concerned on the other. On the basis of such housing demand, research and cross-sectional analysis can be carried out, local, regional, and national policy can be formulated for diverse areas (division of housing space, restructuring, housing costs, new-build supply). The data gathered in a one-off housing demand study can be used to more accurately specify the housing supply (market segments) and demand (consumer groups), and to determine the current and desired division of housing space and the discrepancy between the two. There is, however, no reference point.

For this reason, the questionnaire is often repeated in other housing market areas (or a national study split up into regions) or at another time in the same area. Using the same questionnaire means that the findings can be compared directly and the relative value of the results can be determined.

Cross-sectional data gain more value, however, when the survey is repeated in the same study area at regular intervals and with the same measurement instrument. In this way, a whole series of information is created at an aggregated level about unequivocal operationalized variables. It is then possible to determine changes over time for individual variables and also relations between the variables. Possible trends can then be identified. The Housing Demand Survey (1977–2002) and its successor Housing Research Netherlands (2006–2009) provide the opportunity to conduct a repeated cross-sectional analysis.

In the Dutch surveys referred to the actual relocation movements are determined, which means that changes in the demand for dwellings (in a particular market segment) can be mapped out. This change in demand can then be related to other macro changes, which may provide an explanation for the change.

The study by Boumeester (2004) is used in this chapter as an example of such an analysis. The change in the number of households that move to a detached or semidetached house can be mostly explained by some housing market factors and social-economic and demographic variables. The model estimated for the period 1975–1997 can then be used to make predictions for changes in the demand for detached and semidetached houses between 1998 and 2007.

2.7 Appendix

Table A.1 Data used in the time-series analyses of the realized demand for (semi)detached owner-occupied dwellings

Period	Demand	Real price	Interest rate	Supply	Aged 65+	Nominal price
1973-1		162,100	8.50	13,312	13,672	6.60
1973-2		188,100	8.66	14,445	14,592	10.00
1974-1		178,600	8.59	15,719	15,511	8.90
1974-2		220,000	9.65	17,138	16,038	9.80
1975-1		192,400	9.95	15,039	16,565	6.60
1975-2	0.53	208,000	9.40	12,700	15,966	3.00
1976-1	0.53	221,700	8.87	11,501	15,368	12.30
1976-2	0.52	252,800	8.79	10,165	13,723	14.80
1977-1	0.58	297,000	8.89	11,162	12,079	19.90
1977-2	0.65	321,300	8.67	13,255	14,793	16.00
1978-1	0.69	330,700	8.41	12,630	17,506	7.50
1978-2	0.75	314,400	8.29	14,760	16,646	0.90
1979-1	0.71	300,000	8.85	16,797	15,785	−2.70
1979-2	0.71	279,700	9.55	13,985	16,500	−3.10
1980-1	0.72	259,300	10.64	13,859	17,214	−3.80
1980-2	0.66	236,700	11.36	13,870	15,278	−4.80

(continued)

Table A.1 (continued)

Period	Demand	Real price	Interest rate	Supply	Aged 65+	Nominal price
1981-1	0.62	219,800	11.58	12,013	13,342	−4.90
1981-2	0.55	193,600	12.28	8,420	13,234	−6.30
1982-1	0.46	179,400	11.74	5,672	13,127	−6.90
1982-2	0.37	175,000	10.22	4,040	11,527	−2.90
1983-1	0.38	178,400	9.09	2,867	9,927	1.00
1983-2	0.44	174,600	8.72	2,960	10,049	1.40
1984-1	0.39	169,600	8.62	2,394	10,172	−0.80
1984-2	0.43	168,900	8.35	2,315	10,408	−0.30
1985-1	0.42	166,300	8.20	2,914	10,644	0.30
1985-2	0.48	166,400	7.84	2,910	15,183	0.10
1986-1	0.54	172,700	7.34	3,587	19,721	2.00
1986-2	0.57	176,700	6.98	4,540	18,567	2.90
1987-1	0.57	183,200	6.90	5,996	17,413	2.60
1987-2	0.54	183,500	6.98	8,040	17,720	2.10
1988-1	0.55	188,400	6.85	10,504	18,027	2.10
1988-2	0.59	191,700	6.87	12,340	18,248	2.70
1989-1	0.59	198,900	7.49	12,874	18,470	3.10
1989-2	0.60	202,700	8.18	14,415	16,389	3.40
1990-1	0.63	202,200	8.98	15,444	14,309	1.90
1990-2	0.57	195,400	9.65	15,675	14,159	−0.50
1991-1	0.52	196,700	9.58	16,182	14,009	0.40
1991-2	0.54	198,800	9.49	14,910	13,518	3.00
1992-1	0.53	205,000	9.37	14,950	13,027	4.00
1992-2	0.51	208,900	9.13	17,120	12,971	3.90
1993-1	0.54	215,800	8.54	18,539	12,916	3.80
1993-2	0.60	223,100	7.51	20,025	12,075	4.80
1994-1	0.59	232,400	7.02	19,489	11,234	5.30
1994-2	0.56	229,700	7.56	23,416	12,012	2.80
1995-1	0.59	231,700	7.98	24,124	12,791	1.00
1995-2	0.62	239,800	7.40	25,530	13,220	3.00
1996-1	0.63	250,600	6.64	23,615	13,649	5.00
1996-2	0.63	253,800	6.25	19,851	12,566	4.00
1997-1	0.66	266,200	6.02	19,649	11,483	4.50
1997-2	0.69	272,100	5.93	20,502	12,212	4.60

Source: WBO 1977/1978, 1981, 1985/1986, 1989/1990, 1993/94, 1997/1998 and Statistics Netherlands 2004 (OTB/TUDelft calculations)

Demand number of movers to (semi-)detached owner-occupied dwellings, as a percentage of the total number of households, *real price* average real house prices (corrected for inflation, in terms of 1994s prices), *interest rate* average nominal mortgage interest rate, *supply* number of newly built dwellings in the middle and upper price classes, on a half year basis (a two periods average), *aged 65+* increase (absolute) of the number of persons in the age category 65 years or older, on a yearly basis (a two periods average), *nominal price* changes (%) of the average nominal house prices, on a half year basis (a two periods average)

Table A.2 Explanatory model for the level of actual demand for detached and semi-detached owner-occupied dwellings in the period 1975–1997

Model profile[a]	*b*	*t*	*B*
Model B			
Constant	0.227	(4.29)	
Real price{1}	0.002	(12.27)	0.771
Interest rate{2}	−0.023	(−5.93)	−0.354
Supply {1}	0.469 E-05	(5.43)	0.332
Aged 65+	0.010 E-03	(5.59)	0.292
Nominal price2{3}	−0.004	(−3.67)	−0.230
Dummy93{3}	−0.047	(−2.8)	−0.193
N	45	Sign op 5%	
R^2	0.931		
SEE	0.024		
Durbin–Watson	1.844		

Source: WBO 1977/1978, 1981, 1985/1986, 1989/1990, 1993/94, 1997/1998 (OTB/TUDelft calculations)

Real price average real house prices (corrected for inflation, in terms of 1994s prices), *Interest rate* average nominal mortgage interest rate, *Supply* number of newly built dwellings in the middle and upper price classes, on a half-year basis (a two periods average), *Aged 65+* increase (absolute) of the number of persons in the age category 65 years or older, on a yearly basis (a two periods average), *Nominal price* changes (%) of the average nominal house prices, on a half-year basis (a two periods average)

[a] The addition {1} means that the independent variable effects the dependent variable with a lag of one period (half year)

Table A.3 Data used as input for the three scenarios in the prediction of the realized demand for (semi)detached owner-occupied dwellings, in the period 1998–2007

		Real price			Nominal price			Interest rate			Supply		
Period	Aged 65+	A	B	C	A	B	C	A	B	C	A	B	C
1998-1	12,940	281,300			3.4			5.73			21,788		
1998-2	11,774	296,100			5.3			5.40			22,980		
1999-1	10,608	318,600			7.6			5.01			22,308		
1999-2	10,681	338,900			8.2			5.28			21,335		
2000-1	10,754	362,900			7.9			5.65			20,204		
2000-2	10,892	375,500			6.8			6.11			19,682		
2001-1	11,030	385,800			5.4			5.95			20,008		
2001-2	11,568	389,600			4.1			5.64			20,899		
2002-1	12,107	394,200			2.9			5.43			20,442		
2002-2	12,886	395,200			2.4			5.24			19,188		
2003-1	13,665	395,200	394,200	396,200	1.3	1.2	1.4	5.30	5.30	5.17	20,000	21,000	21,000
2003-2	14,965	395,200	390,500	396,200	1.1	0.5	1.2	5.30	5.30	5.09	20,000	19,000	19,000
2004-1	16,266	395,200	389,500	404,900	0.8	0.2	1.9	5.30	5.80	5.02	20,000	18,500	23,500
2004-2	16,731	395,200	380,800	411,900	0.5	−0.8	2.5	5.30	6.30	4.94	20,000	16,500	21,500
2005-1	17,197	395,200	375,600	422,100	0.6	−1.2	2.7	5.30	6.80	4.87	20,000	16,000	26,000
2005-2	18,075	395,200	364,300	430,000	0.8	−1.4	3.0	5.30	7.30	4.80	20,000	14,000	24,000
2006-1	18,953	395,200	353,200	442,900	0.8	−2.3	3.2	5.30	7.80	4.72	20,000	13,500	28,500
2006-2	17,432	395,200	337,.500	453,100	0.8	−3.0	3.4	5.30	8.30	4.65	20,000	11,500	26,500
2007-1	15,911	395,200	323,300	469,100	0.8	−3.6	3.7	5.30	8.80	4.57	20,000	11,000	31,000
2007-2	14,400	395,200	306,100	481,900	0.8	−4.0	3.9	5.30	9.30	4.50	20,000	9,000	29,000

Real price average real house prices (corrected for inflation, in terms of 1994s prices), *Interest rate* average nominal mortgage interest rate, *Supply* number of newly built dwellings in the middle and upper price classes, on a half-year basis (a two periods average), *Aged 65+* increase (absolute) of the number of persons in the age category 65 years or older, on a yearly basis (a two periods average), *Nominal price* changes (%) of the average nominal house prices, on a half-year basis (a two periods average)

References

Berry, W. D., & Feldman, S. (1985). *Multiple regression in practice. Quantitative applications in the social sciences* (Vol. 50). Newbury Park: Sage.

Boumeester, H. (2004). Duurdere koopwoning en wooncarrière. Een modelmatige analyse van de vraagontwikkeling aan de bovenkant van de Nederlandse koopwoningmarkt. *Volkshuisvestingsbeleid en woningmarkt, 35*. Delft: Delft University Press.

Boumeester, H., & van der Heijden, H. (2004). *Marktimperfecties, conjunctuurgevoeligheid en segmenten op de woning(bouw)markt*. Delft: Onderzoeksinstituut OTB.

Boumeester, H. J. F. M., Lamain, C. J. M., Marien, A. A. A., Nuss, F. A. H., & Rietdijk, N. (2004). *Huizenkopers in profiel. Onderzoek naar wensen van potentiële huizenkopers*. Voorburg: NVB.

Boumeester, H., Hoekstra, J., Meesters, J., & Coolen, H. (2005). *Woonwensen nader in kaart: de woonbeleving van bewoners*. Voorburg: NVB.

Boumeester, H. J. F. M., Lamain, C. J. M., Marien, A. A. A., Nuss, F. A. H., & Rietdijk, N. (2008). *Huizenkopers in profiel. Onderzoek naar wensen van potentiële huizenkopers*. Voorburg: NVB.

Boumeester, H., Dol, K., & Meesters, J. (2009). *Stedelijk wonen; een brug tussen wens en werkelijkheid. Een onderzoek naar woonwensen en woonproducten bij binnenstedelijk bouwen*. Voorburg: NVB.

Clark, W. A. V., & Dieleman, F. M. (1996). *Choice and outcomes in the housing market*. New Brunswick, NJ: Centre for Urban Policy Research.

Clark, W. A. V., Deurloo, M. C., & Dieleman, F. M. (1990). Household characteristics and tenure choice in the US housing market. *Netherlands Journal of Housing and Environmental Research, 5*(3), 251–270.

Cohen, J., Cohen, P., West, S. G., & Aiken, L. S. (2003). *Applied multiple regression/correlation analysis for the behavioral sciences* (3rd ed.). London: Lawrence Erlbaum.

Coleman, J. S. (1990). *Foundations of social theory*. Cambridge, MA: The Bellknap Press of Harvard University Press.

de Groot, C., Manting, D., & Boschman, S. (2008). *Verhuiswensen en verhuisgedrag in Nederland. Een landsdekkend onderzoek*. Den Haag: PBL.

Draper, N., & Smith, H. (1981). *Applied regression analysis* (2nd ed.). Toronto: Wiley.

Elsinga, M. (1995). Een eigen huis voor een smalle beurs: het ideaal voor bewoner en overheid? (Dissertation, *Volkshuisvestingsbeleid en bouwmarkt, 23*, Delftse Universitaire Pers, Delft).

Elsinga, M., & Goetgeluk, R. (2003). Onderzoek woningbouwprogramma De Volgerlanden. Visie op Vinex-locatie Hendrik-Ido Ambacht. OTBouwstenen 83, Delft: Delft University Press.

Everaers, P. C. (1990). *Residential mobility in the Netherlands: a descriptive analysis based on the Housing Demand Survey 1985/1986*. Supplement bij de sociaal-economische maandstatistiek, pp. 28–45

Field, A. (2004). *Discovering statistics using SPSS for Windows*. London: Sage.

Goetgeluk, R. (1997). *Bomen over wonen. Woningmarktonderzoek met beslissingsbomen* (Dissertation, Faculteit Ruimtelijke Wetenschappen Universiteit Utrecht).

Goetgeluk, R. W., Hooimeijer, P., & Dieleman, F. M. (1992). *The effectiveness of housing search: the role of motives for moving and housing market adjustment*. Paper presented on the International Research Conference: European Cities, Growth and Decline, Den Haag.

Heins, S. (2002). *Rurale woonmilieus in stad en land, plattelandsbeelden, vraag en aanbod van rurale woonmilieus*. Delft: Eburon.

Hilkhuysen, G. (1999). *Enquêteren op het OTB: een systeembeschrijving*. Delft: Delft University Press.

Hooimeijer, P., & Linde, M. (1988). Vergrijzing, individualisering en de woningmarkt (Dissertation, Elinkwijk bv, Utrecht).

Kruythoff, H. (1993). Residential environments and households in the Randstad (Thesis, Housing and Urban Policy Studies, 8, Delft University Press)

Kulberg, J., Ras, M. (2004). *Met zorg gekozen? Woonvoorkeuren en woningmarktgedrag van ouderen en mensen met lichamelijke beperkingen*. Sociaal en Cultureel Planbureau en Ministerie van VROM, Den Haag (Ministerie van VROM).

Laakso, S., & Loikkanen, H. A. (1992). *Finnish homes; through passages or traps? An empirical study of residential mobility and housing choice*. Paper presented on the International Research Conference: European Cities, Growth and Decline, Den Haag.
Lewis-Beck, M. S. (1980). *Applied regression. An introduction* (Quantitative applications in the social sciences, Vol. 22). Newbury Park, CA: Sage.
Lindenberg, S. (1990). Homo socio-oeconomicus: The emergence of a general model of man in the social sciences. *Journal of Institutional and Theoretical Economics, 146*, 727–748.
Maddala, G. S., & Lahiri, K. (2009). *Introduction to econometrics*. Chichester: Wiley.
Moore, E. G., & Clark, W. A. V. (1990). Housing and households in American cities:structure and change in population mobility, 1974–1982. In D. Myers (Ed.), *Housing Demography*. Madison, WI: University of Wisconsin Press.
Mulder, C. H. (1993). *Migration dynamics: a life course approach (thesis)*. Amsterdam: Thesis Publishers.
Mulder, C. H., & Hooimeijer, P. (1995). Moving into owner-occupation: Compositional and contextual effects on the propensity to become a home-owner. *Netherlands journal of Housing and the Built Environment, 10*(1), 5–25.
Oskamp, A. (1997). *Local housing market simulation; a micro approach (thesis)*. Amsterdam: Thesis Publishers.
Priemus, H. (1969). *Wonen; kreativiteit en aanpassingen; onderzoek naar voorwaarden voor optimale aanpassingsmogelijkheden in de woningbouw*. Den Haag: Mouton & Co.
Priemus, H. (1984). *Verhuistheorieën en de verdeling van de woningvoorraad*. Delft: Delftse Universitaire Pers.
Teule, R. (1996). *Inkomen, doorstromen en uitsorteren: arm en rijk op de Nederlandse grootstedelijke woningmarkt*. Thesis, Delfse Universitaire Pers.
VROM. (2001). *Nota mensen, wensen wonen*. Den Haag: Ministerie van VROM.
VROM. (2002). *Beter thuis in wonen. Kernpublicatie WoningBehoefte Onderzoek 2002*. Den Haag: Ministerie van VROM.
VROM. (2003). *Beter thuis in wonen. Kernpublicatie WoningBehoefte Onderzoek 2002*. Den Haag: Ministerie van VROM.
VROM. (2007a). *Dynamiek in de derde leeftijd. De consequenties voor het woonbeleid*. Den Haag: Ministerie van VROM.
VROM. (2007b). *Bouwen voor de schuifpuzzel*. Den Haag: Ministerie van VROM.
VROM. (2007c). *Wonen op een rijtje. De resultaten van het WoonOnderzoek Nederland 2006*. Den Haag: Ministerie van VROM.
Willekens, F. (1989). *Understanding the interdependence between parallel careers*. Paper presented at the workshop "Female labour market behaviour and fertility: preferences, restrictions, behaviour". Den Haag: NIDI.

Chapter 3
The Decision Plan Nets Method

Roland W. Goetgeluk

3.1 Introduction

In this chapter we explain how and why the decision plan nets method (DPN), also known as Decision Net or Decision Tree (DT), has been used in fundamental and applied research to describe, explain, and predict how people trade off housing preferences if supply fails (Bettman 1979; Park et al. 1981; Timmermans and Van der Heijden 1987; Op 't Veld et al. 1992; Van Kempen et al. 1994; Van Zwetselaar and Goetgeluk 1994; Floor et al. 1996; Louwers 1996; Goetgeluk et al. 1995; Goetgeluk 1997; Witlox 1995; Wets 1998, Arentze & Timmermans 2007, Heins 2002; Goetgeluk and Hooimeijer 2002; Daalhuizen 2004; Boumeester et al. 2008).

In this chapter, we will discuss the advantages and disadvantages of the DPN. The chapter is organized as follows. In Sect. 3.2, we discuss the value of the DPN within the theoretical background of housing choice. In Sect. 3.3, we discuss the main technical aspects of the DPN and specifically the interview. Section 3.4 presents examples. The last section draws some conclusions and shows prospects.

3.2 Housing Demand, Housing Supply, and the Trading off of Housing Preferences

In our view the study of housing choice must answer the question of how individual housing choice can be better described, explained, and simulated, given a meaningful relationship between individual housing choice and the housing market as a system in which supply and demand intersect. Central to this discussion is the question of

R.W. Goetgeluk (✉)
Demography & Housing, ABF Research, Delft, The Netherlands
e-mail: roland.goetgeluk@abf.nl

S.J.T. Jansen et al. (eds.), *The Measurement and Analysis of Housing Preference and Choice*, DOI 10.1007/978-90-481-8894-9_3,

how macro developments such as a lack of housing supply affect micro behavior, such as the housing choice of individuals, and vice versa.

A conceptual model of this housing choice process is available (Hooimeijer and Linde 1988; Mulder 1993; Goetgeluk et al. 1995; Mulder and Hooimeijer 1999). In this model, individual housing choice not only depends on the supply within the housing market, but also affects the supply. What characterizes the housing market is that supply – regarded as a characteristic of the macro level – is generated for the most part by people moving house (filtering supply). This means that supply is in part a side effect of the choices made by housing consumers. The extent to which supply is made available and is reoccupied by various categories of households could be in line with government policy or may run counter to it. For that reason, the government seeks to regulate the housing market through legislation, which affects the individual household's access to the supply.

At the individual or household level, the supply must be linked to the preferences of the household. According to the conceptual model, the housing consumer will learn how the housing market enables or constrains the housing search. These opportunities and constraints occur at both the micro and the macro level. At the individual level, the opportunities are the household's resources such as income, while the constraints include accessibility to the place of work or the children's school. Time constraints may also apply, because usually the accommodation must be found within a specific period. The degree of that constraint depends on the urgency of the move (motive for moving). At the macro level, the availability and accessibility of the supply constrains choices. It is safe to say that after a period of searching, people realize that their original preferences were unrealistic.

The matching of supply and demand can be represented in housing market simulation models. The present body of research provides a good insight into the demand for accommodation (motivational aspect) and the development of supply (situational aspect). However, modeling how an individual's preferences are matched to specific vacancies is problematic (procedural aspect). We need methods to show how the willingness of people to accept a particular dwelling can be modeled, given the variation in the urgency of the move, in household resources and preferences and in opportunities and constraints at the macro level (Goetgeluk 1997; Oskamp 1997).

There are therefore consequences for how the choices people make are measured. Housing market research uses two kinds of models to describe, explain, and predict the choices: revealed choice and stated choice models. Revealed choice models analyze moves that have taken place. The manifest (overt) choice is equated with the housing preferences of the housing consumer. This does not explain how housing consumers respond to changes in the supply. Another approach is to analyze housing preferences, which are not always manifest in a move, using stated choice models. Rather than quantifying house-moving patterns, these models quantify how alternatives are evaluated. The stated choice method fails if no explicit conditions are created in which the housing consumers must make choices when there is a lack of supply.

In that case, one possibility is to use an algebraic stated choice model. However, this is not always an option because most models are additive. In an additive model, a low score (say, zero) on a single attribute of the dwelling does not necessarily lead to a poor overall valuation and rejection of the alternative. In other words, models of this type are fully compensatory. The possibility that a house could be rejected based on just one preference is not taken into account. It can safely be assumed that some preferences will be less compelling than others, and these can also vary from one housing consumer to the next. Sometimes a preference is rigid, sometimes open to substitution, and sometimes flexible. If a rigid preference is not met, the dwelling will be rejected. A substitutable preference is one that does not lead to rejection, but must be compensated by some other preference. Flexible preferences never lead to refusal, but they do make one house more attractive than another. An alternative model is required to reveal the structure of trade-offs between housing preferences. In this chapter the DPN is used as this alternative

3.3 DPN As an Alternative

Bettman (1979) describes a DPN as "another alternative for the representation of consumer choice heuristics. In the branching structure attributes [price] and situational factors can be used to predict the acceptance or rejection of an alternative." Bettman perceives the housing consumer as a processor of information. In that vein, the key concepts in the search process are the motivation, information acquisition, and decision-making. His approach is therefore also focused on the motivational, situational and procedural aspects of housing choice!

The DPN is based on a structured (computerized) interview that shows people's choice processes. It reveals a set of imaginary houses that the housing consumer would consider acceptable based on a mix of compensatory and non-compensatory choice rules that respondents are free to mention. It has the structure of a tree or net. We use the tree. Like a tree, we make a distinction between the main branch and the lateral branches (Fig. 3.1). The DPN is literally depicted and reveals the attributes (rooms, garden, garage, attic), the operators (>, =, etc.), the levels (4, 5, YES, NO), and the choice rules.

The interview has two steps: revealing the main branch and the lateral branches.

1. The interview starts by asking a consumer which attributes will be considered. It is vital that a saturated set of attributes is known since this is required for the second step. Every time an attribute is mentioned, we depict it as a node on the main branch. In Fig. 3.1 the main branch consists of three nodes: "Rooms >5 AND Garden=yes AND Garage=yes." The sum depicts an imaginary house and the most preferred.
2. The interviewer triggers the consumer to reveal the choice rules that are used if a vacancy that fails to satisfy only one of the attributes of the main branch is offered. Since we have three nodes, we have three-pronged questions, for

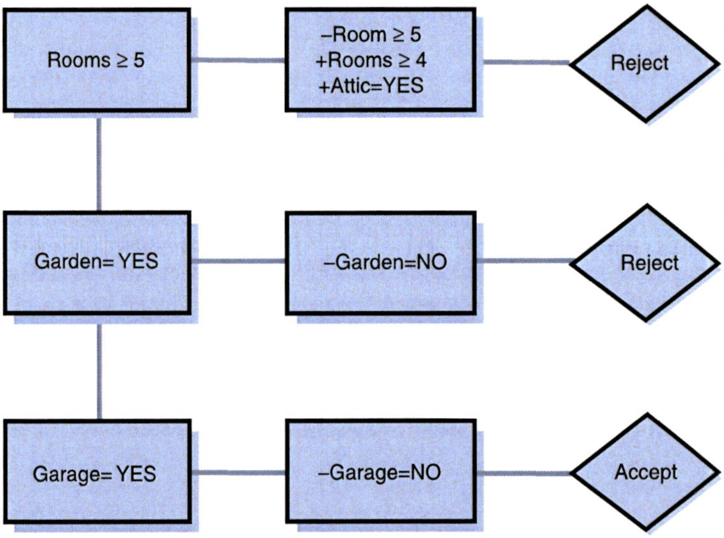

Fig. 3.1 A DPN (After Van Zwetselaar and Goetgeluk 1994)

example: "Would you consider a vacancy that has a garden and a garage, but the number of rooms is five or more?" We stress that it is vital that all other main branch attributes remain constant! A deviation from this rule invokes measurement errors that can only partially be corrected afterwards (Van Zwetselaar and Goetgeluk (1994). The answer reveals the choice rules. They can only be threefold. Park et al. (1981) link the rules to the attributes and call them dimensions: rejection-inducing dimensions, relative-preference dimensions and trade-off dimensions. We will use an adapted set of abbreviations in the remainder of this chapter, and we will use the commonly used word "preference" instead of "dimension" – rejection-inducing preference (RIP), relative-preference (RP), and trade-off preference (TOP):

- An RIP is a preferred attribute, which, if not present, will lead to immediate rejection of the offered dwelling. An RIP is rigid or a claim, like the availability of a garden. If an RIP results in a "reject" no branching will occur at that point.
- The RP leads to the ranking of alternatives. The housing consumer will give a higher score to an alternative that has a preferred attribute than one that lacks it. This resembles the example presented above. The Garage is an RP in Fig. 3.1.
- A TOP emerges when a preferred attribute is not satisfied, but other attributes can compensate for it. The preference Rooms is a TOP. Fewer rooms may be compensated for by, for example, an attic. An attic is often mentioned as a less valuable kind of room since it is often less usable than a normal room. The TOP leads to lateral branching and must end with a "reject." "Would you consider a vacancy that has a garden and a garage, but the number of rooms is five or

more?" has to be followed by a new question: "Would you consider a vacancy that has a garden and a garage, but the number of rooms is four or more and an attic?" In this case, the consumer rejects the offer. We stress again that this compensation may never be one of the remaining attributes on the main branch. The lateral branch can be very long since the attic may be a TOP as well and might be compensated by, for instance, another attribute like "storage room in garden." Important to know is that the last node of the lateral branch by definition is an RIP. In noncomputerized versions of the DPN violating this rule results in measurement errors (Van Zwetselaar and Goetgeluk 1994).

We conclude that this method (using the DPN) differs from those methods where the number of attributes, their levels, and the combination of attribute levels are predetermined, such as in conjoint models. The DPN also allows for a mix of compensatory and non-compensatory rules to be used.

3.4 Applications

3.4.1 Explorative Research

Often the DPN is used in the explanatory phases of research to determine which preferences are important to people. Based on frequency distribution the commonly mentioned preferences and their values are selected.

Two examples illustrate the strength of the DPN in explorative research. The studies by Heins et al. (2003) and Boumeester et al. (2008) show how the DPN reveals unknown attributes. The study by Boumeester et al. (2008) also showed the importance of checking assumptions with regard to attribute levels.

Heins et al. (2003) used the simplified DPN to analyze the preferences for rural housing in the Netherlands. Corresponding to the conceptual framework described in Sect. 3.2, they analyzed the preferences of respondents in two cities and two suburban municipalities in a housing market region in the Randstad-Holland (Utrecht) and a southern province Noord-Brabant (Den Bosch). Both regions differ with respect to the pressure on the housing market and in the amount of private and public green space. They wanted to test whether "rural housing" refers to being housed in the countryside or to an image or brand that can also be located in urban areas. Table 3.1 shows one of the results of this study.

Table 3.2 shows the preferred attributes that were mentioned by at least 68% of the respondents ($n = 112$). They used two steps to unravel all the information in the table. First, they defined the preferred attribute by type. They combined house (H), location (L), function (F) and esthetics (A). For instance, HF refers to a functionally preferred attribute of a house like Garden/Balcony (column type). In the second step, they calculated the share of each preferred attribute (column % total). For instance, "Garden/balcony" is mentioned by 98 of each 100 respondents. The minimum in the table was "presence of animals," which was shared by 68 of each 100 respondents. Heins et al. (2003) argued that the RP is less important if we want to

Table 3.1 Most frequently stated preferred attributes by type and ranked by weighted share (Heins et al. 2003)

Preferred attribute	Type	% Total	% RIP of Total	% TOP of Total	% RP of Total	((% RIP + % TOD) × % Total)/100
Garden/balcony	HF	98	82	15	4	95
Location	LF	95	66	28	6	89
Green space	LF/LA	95	65	25	9	86
Number of rooms	HF	95	80	11	9	86
Type of dwelling	HF	92	47	47	7	86
Personal safety	LF	91	88	3	9	83
Peace and quiet	LA	90	88	4	8	83
Density	LA	90	70	22	8	83
Proximity to shop	LF	87	90	5	5	83
Type of buildings	LA	92	56	33	11	82
Plot size	HF	84	75	15	11	76
Type of landscape	LA	89	47	37	16	75
Road safety	LF	83	76	10	14	71
Outbuildings	HF	89	58	22	20	71
Architecture – surroundings	LA	84	71	12	17	70
Atmosphere	LA	90	69	7	24	68
Open space	LF/LA	86	71	8	21	68
Proximity to nature	LF/LA	87	47	23	30	61
Architecture – dwelling	HA	71	71	13	17	60
Type of population	LA	77	62	14	24	59
Age of building	LA	62	48	26	26	46
Proximity to water	LF/LA	77	27	24	49	39
Presence of animals	LF/LA	68	38	8	53	31

H house, *L* location, *F* function, *A* esthetics

gain insight into what they define as the "rural" profile or housing bundle. The table shows that in 94 of the 100 respondents the preferred attribute "Garden/Balcony" is an RIP or a TOP. However, the preferred attribute "Presence of animals" has a significantly lower percentage (31). If we use this percentage as a weight factor, the ranking of Table 3.2 becomes: ((RIP + TOP)/100)* Total.

It turns out that most preferred attributes are linked to functionality except for many of the preferred attributes with respect to rural housing (HA and LA). The preferred attributes linked to the location are more often defined by the researchers as esthetics (Heins 2002). Other aspects relate to the way people act in a neighborhood (type of population), the maintenance of public space (safety) or a "feeling" that was attached to a preferred attribute. We can also define these as the social aspects. This would result in the subdivision of the location features into function (amenities), social (people), and design (esthetics).

The weighted ranking shows that housing and location-related preferred attributes were mixed in the upper part of the ranking, while lower down location dominates.

Table 3.2 The original classes for the conjoint model and the mean preferred class of the DPN (Boumeester et al. 2008)

Attribute	Predefined conjoint model levels	DPN levels	Ratio DPN/CM
Size of living room			
Minimum	20 m^2	31 m^2	1.6
Maximum	40 m^2	48 m^2	1.2
Number of rooms			
Minimum	2	4	2.0
Maximum	4	5	1.3
Rental price			
Minimum	338 €	387 €	1.1
Maximum	725 €	563 €	0.8
Buying price			
Minimum	140,000 €	222,727 €	1.6
Maximum	300,000 €	395,109 €	1.3

This is in concordance with other DPN studies (Goetgeluk 1997; Daalhuizen 2004; Boumeester et al. 2008). What is more interesting is that location-preferred attributes like "green space," "peace and quiet" and "safety" score high. However, preferred attributes like "open space", "proximity to nature," or even "presence of animals" score low. This suggests that "rural" housing does not necessarily have to be associated with agriculture, the agribusiness, and its effects on the landscape.

Based on many more different analyzes with different datasets, Heins (2002) concluded that the countryside is a construct. This implies that the construct or image can be used as a brand in marketing. Furthermore, they argued that "pseudo-countrysides" can be designed and developed within the urban fringe. And indeed, currently, "rural" neighborhoods are developed and marketing uses "buzz words" that appeal to the best that both the countryside and the city have to offer.

Boumeester et al. (2008) used the same procedure as Heins to test whether the attributes and their levels of a conjoint model were justified. Since these profiles were used in a nationwide large-scale representative sample in the Netherlands, any misspecification could lead to enormous financial losses and of course useless findings. We will only show an example of how the DPN is used to test the appropriateness of the category levels used.

The study showed that prior defined attribute levels were not always justified. Table 3.2 shows the results of the predefined minimum and maximum levels of the attributes related to space and tenure/price, and the levels according to the DPNs; in other words, the minimum and maximum preferred size, number of rooms and price, as indicated by the respondents. If the ratio between the DPN and the conjoint model levels fluctuates around 1, then we can conclude that the predefined levels are justified. However, Table 3.2 shows that most minimum levels were wrongly defined. The DPN levels reflected consumers' perceptions of how quality is related to price. This result also improved the conjoint model since the realistic levels of the attribute define the outcome of the conjoint model. This can be illustrated by the size of the living room. Therefore, the result of the conjoint models depends on a valid set of attributes and a valid set of attribute levels.

Both examples show the value of the DPN as a fairly simple method to analyze at an aggregate level the character of preferences or to determine which attributes and their levels should be used in other more restricted preference models.

3.5 Preference Modeling and Knowledge-Based Simulation Models

Figure 3.1 showed that the DPN method is not rooted in algebraic mathematics, like the conjoint model or hedonic price model, but in "Set theory" (Timmermans et al. 1994, Van Zwetselaar and Goetgeluk 1994; Witlox 1995; Wets 1998). This theory states that an object, such as a real-world vacancy (supply), can be a member of a set of objects, such as a set of imaginary houses that the housing consumer would consider acceptable as well as the rules that determine when they consider them. The set of imaginary houses and rules is the DPN (demand). If we look again at the DPN above in Fig. 3.1, we see that it is a set of objects in which each object is an exhaustive set of independent choice rules that lead to an event to accept. For instance, the independent choices rules, ranked by utility, to accept a dwelling of Fig. 3.1 are:

IF Offer is
element of Set {{Garden = Yes} AND {Rooms >5} AND {Garage = Yes}}
Then Accept
OR
element of Set {{Garden = Yes} AND {Rooms>4 AND Attic = Yes} AND {Garage = Yes}}
Then Accept
OR
element of Set {{Garden = Yes} AND {Rooms >5}}
Then Accept
End If

In research, we use a representative survey and have a large set of choice rules. If the choice rules have predictive power, which can be tested in, for instance, a longitudinal survey (see below), we may use the DPNs in a decision support system (DSS). The DPN is a substitute for a respondent. The DSS accepts all offers, which are defined as a set of attributes and their levels. A DPN (respondent) evaluates an offer, which must result in a reject or accept. The final result is a sum of rejects and accepts that can be linked directly to the set of attributes and their levels of each offer.

The use of this property of the DSS is twofold. The DSS is an interface that allows researchers, policy-makers, real estate agents, project developers, or housing corporations to offer real supply to the DPNs. It allows them to change attribute levels like price for sensitivity analysis. Van Zwetselaar and Goetgeluk (1994) developed such a DSS "housing supply machine."

A necessary condition to use the DSS is the predictive value of the DPN. Goetgeluk's study "Trading off housing preferences; housing market research with Decision Plan Nets" analyzed the DPNs predictive value in a longitudinal perspective (1993–1994). The respondents originated from a representative (longitudinal) sample of active searchers in two Dutch housing market regions, Utrecht and Arnhem. These regions were chosen based on an earlier analysis that had shown that demand exceeded supply far more in the Utrecht region than in the Arnhem region. A housing market region is the aggregate search area of housing searchers and is defined as the area in which a person moves without loss of accessibility (time or social distance) to work, family, and friends. The selection of a housing market region therefore implies that the motives "improve housing" and "household formation" overrule "study/work" as was indicated in Sect. 3.2. The analysis consisted of three stages.

The first stage of the analysis took a closer look at the preference structure of various housing consumers. In Sect. 3.2 it was argued that valid measurements of stated preferences can only be done within the framework of a meaningful relationship between individual housing choice and the housing market as a system in which supply and demand intersect. Indeed, it turned out that if the urgency of the move, individual resources and the search area were taken into account, the housing consumers had different sets of preference. The differences lie not so much in the type of preferred attributes, but in the number and the rigidity. Housing consumers whose move is not too urgent and who have ample resources provide more RIPs and relatively few TOPs. This was predicted.

In the second stage, it was tested if the stratification based on stage 1 resulted in differences between the various respondents in the propensity to move within 1 year. It was observed that the propensity to move house differed considerably across various categories of housing consumers, in accordance with expectations of the conceptual framework of Sect. 3.2. Especially the urgency of the move proved to be significantly important: the more urgent the move, the higher the propensity. On an average, almost half of the actively searching housing consumers moved within a year. This share was predicted.

The third stage examined the proportion of movers who had accepted a house that matches one of the alternatives specified in the DPN. This is expressed as the success rate. The success rate is therefore dependent on the propensity to move. A calculation was also made of the extent to which the preferences of housing consumers still searching in 1994 differed from the preferences they expressed in 1993. The success rate was 0.25. It was shown that the shift in preferences of those still looking for a house was many times greater than the difference between the choice of house and the original preferences stated by people who had already moved. This suggests that people looking for houses adapt their preferences constantly, especially if the urgency to move is less high! This implies that besides our stratification variables "urgency to move" in combination with "would you accept a preferred offer immediately," we have to add the search period as well. Luckily enough one attribute is responsible for the major deviance in time, conditional on the urgency to move: the substitution of price. If we were to take price developments

Table 3.3 Description of segments (Goetgeluk 1997)

Code segment	Description
1A	Starter, single
1B	Starter, couple
2	Mover, low income, <45 age, single
3	Mover, low income, <45 age, couple/single parent
4–7	Mover, low income, couple with children
5	Mover, low income, ≥45 age, single
6	Mover, low income, ≥45 age, couple/single parent
8	Mover, high income, <45 age, single
9	Mover, high income, <45 age, couple/single parent
10–13	Mover, high income, couple with children
11	Mover, high income, ≥45 age, single
12	Mover, high income, ≥45 age, couple/single parent

into account, though, the success rate would be greatly increased. It turned out that substitution of the other preferences like housing type, volume, and size of the neighborhood were fairly stable.

Based on these outcomes, Goetgeluk (1997) and Goetgeluk and Hooimeijer (2002) assumed that the predictive value was sufficient enough to use the DPNs in a knowledge-based DSS. Table 3.3 shows the segmentation of demand according to the framework described in Sect. 3.2. Two kinds of simulations with the DSS are presented here.

3.5.1 *Unravel the Way One Attribute of a Vacancy is Valued, Controlling for All Satisfied Preferences*

In the experiment, we increased the price of a vacancy (ownership). The minimum level started at 0–75,000 Dutch Florins and ended with 526,000 Dutch Florins or more. We increased the level by 50,000 Dutch Florins. We expected a negative nonlinear (S-curved) relationship. We assumed threshold levels: at a certain price level the acceptance rate will remain constant.

Figure 3.2 shows that the relation is as expected. Threshold levels exist and they are segment-specific. The highest acceptance rates were movers with a high income. In all cases, these were couples with children and childless couples younger than 45 years (9, 10, 12, 13). Only the singles deviated from this general view of the high incomes. A small group is formed by the young singles (8) and starters who planned to cohabit (1B). The lowest rates include all the other low-income groups plus the older singles with a high income (11).

Figure 3.2 shows the acceptance rate as a function of only one dynamic attribute level. Here we state that we could offer the DSS a full fractional design, like in conjoint models, to create "rejects" and "accepts."

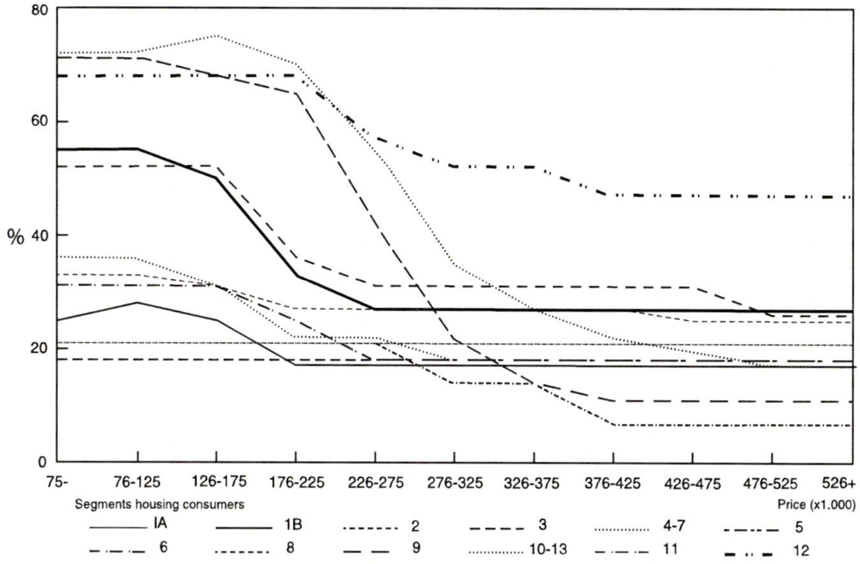

Fig. 3.2 The rate of acceptance for price ownership for various segments of housing consumers (Goetgeluk 1997)

3.5.2 *The Role of Location*

How important is location in the residential choice? The statement "location, location, location" postulates that only location counts. We have seen that location is a latent variable that can be deconstructed in functional, esthetical and social attributes. We have also shown that not all location attributes are valued equally important. So, how important is location?

We tested the importance of the attribute municipality. We offered our DPNs 22 real vacancies. They originated from estate agents' announcements in a special local paper. We selected them randomly. We offered them to all respondents irrespective of their segmentation. Here we present the results of the Arnhem housing market region. We are interested in the acceptance rate. The application is of course very applied. Any real estate agent can use the DSS.

How important is the preferred municipality? As indicated earlier, we know that trading off the location at a low spatial scale in the search region – the housing market region – is at stake. In this scenario, we assumed that the acceptance rate drops if the municipality is out of reach. To test this we defined two "supply" simulations. In the first, our respondents – the DPN – evaluated each offer. In the second simulation, we offered the same vacancies, but the DPNs assume that the preference for the municipality is always satisfied. Hence, the difference in the acceptance rate is to be accounted for by a spatial mismatch: the preferred set of municipalities is either an RIP or a TOP. Our developed DSS housing supply machine (Van Zwetselaar and Goetgeluk 1994) is able to perform such simulations.

In Fig. 3.3 both rates are depicted: black (location not realizable), dark gray (location realizable). The conclusion is that in many instances people have fairly strong location preferences for municipalities. However, at some points the difference between the bars is less. The scenario outcome is in concordance with other empirical data that show that most people move within the boundary of their municipality or define a limited number of preferred municipalities. The DSS shows immediately which part of the supply is relatively indifferent to location and which is not. For real estate agents the DSS is therefore a useful tool to optimize the portfolio.

3.5.3 *The Role of the Search Area*

In this scenario, we tested how important the housing market region structures the demand. We use the same 22 offers as before. We also forced the DPNs to accept that their municipal preferences would be satisfied. This implies that we assume that all offers are located in the preferred set of municipalities. This means that we measure to what extent our two groups evaluate offers by dwelling-type characteristics. We assume that the respondents of Utrecht will have a higher acceptance rate since Utrecht housing seekers are accustomed to less quality for a higher price. In contrast to the other scenario, the DSS offers these vacancies first to housing seekers in Arnhem (dark gray bar) and later to housing seekers in Utrecht (light gray).

Figure 3.3 shows that the impact of the housing market region is huge. In nearly all cases, the acceptance rate for the Utrecht seeker is higher than for the Arnhem seekers. Two conclusions can be drawn. We start with an "applied" one and end with a "scientific" one.

Firstly, using this information, real estate agents could expand their market. Why? We have witnessed housing supply–driven migration from the Utrecht region toward the Arnhem region. Commuting from Arnhem to work in Utrecht – in general the Randstad-Holland – is possible. The cost of an increased commute is less than the benefits of cheaper or better housing. Moving is good value for money. Therefore local real estate agents could extend their market area toward housing market regions within commuting distance.

Secondly, the difference in acceptance rates and the notion that housing market regions are dynamic implies that any research design for housing choice must define the functional housing market as a starting point since the search area particularly influences the attribute levels. This means that regional housing market simulations should at least be tested regularly to check that the regions are still valid. A better solution is an endogenous housing search area within these models. How the functional housing market can be calculated in a flexible way is demonstrated elsewhere (Goetgeluk 1997; Goetgeluk and de Jong 2005).

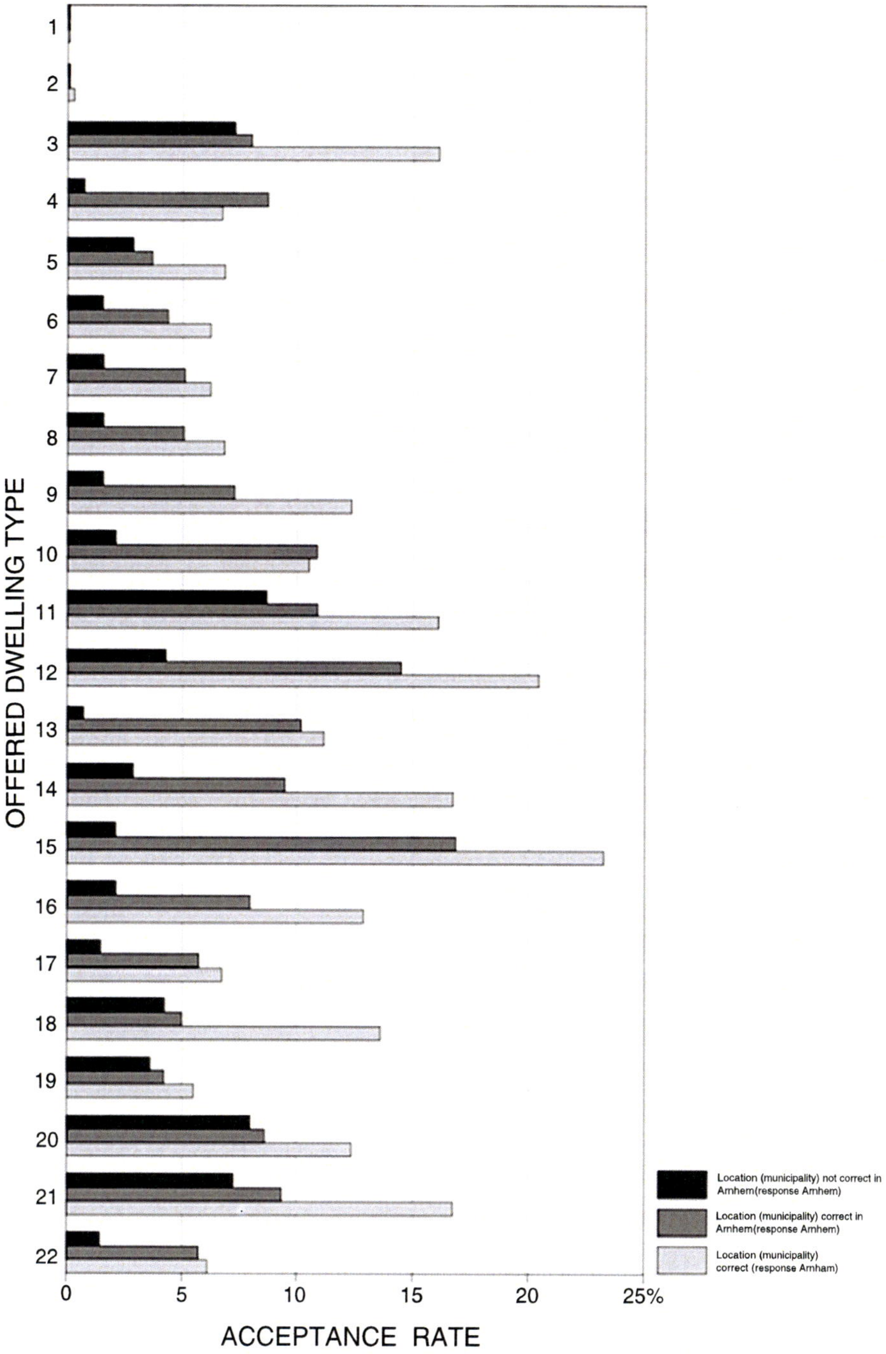

Fig. 3.3 The rate of acceptance of Utrecht and Arnhem respondents for real Arnhem vacancies (Goetgeluk 1997)

3.6 New Prospects

To sum up, the added value of DPNs is twofold. They provide insight into the interdependency of preferences in the evaluation of supply. In addition, they show how people apply their preferences when there is a lack of supply.

Despite the above advantages of the DPN, we nevertheless also found two major drawbacks. The first refers to the problem that the representation of the choice process is not discrete at all. A house with a monthly rent of 601 € is rejected if a maximum rent of 600 € is RIP. This does not make sense as the longitudinal survey showed. Hence a "crisp" DPN seems unjust and a "fuzzy" one better. Contrary to classical set theory, fuzzy set theory states that a house with a monthly rent of 601 € will be acceptable. We must define a probability distribution around the levels. Theoretical and empirical research must show how these intervals are estimated. For a detailed discussion of these problems, we refer to the PhD dissertation of Wets (1998).

The second refers to the explorative value of the DPN. Does this imply that the original DPN is always useful in the explorative phase? Goetgeluk (1997) argues not. Not all conceivable or real options are open to housing consumers. The study "*Ontwerpend aan Holland*" ("Designing the Netherlands"), Geuze et al. 1994 proposes that: "Planners and design disciplines must provide a balance for the unmistakable call from the market for low densities and comfortable living environments. [...] When finally designing the future of the Netherlands in general [...] this design will have to fit in with the day-to-day living patterns of the new residents" (translation Roland Goetgeluk).

In applied research, DPNs are used in a more explorative manner. This has resulted in a "simplified" DPN. It only reveals the main branch and the first node of a lateral branch. Table 3.4 shows the structure of a data matrix in Excel that Louwers (1996), Heins (2002), Daalhuizen (2004) and Boumeester et al. (2008) use. The operator (left, right) and value (left, right) are necessary to take account of the interval values for rent for the first respondent. In reality, the row is much longer: m attributes. To facilitate easy data processing, attributes that are always mentioned or are at stake in the new design are already ordered in this data-matrix and thus also in the (Internet) questionnaire. Related to the second drawback, one aspect of this

Table 3.4 Data structure of a simplified DPN (Louwers 1996; Heins 2002; Daalhuizen 2004; Boumeester et al. 2008)

Respondent	Attribute name 1	Operator left	Value left	Operator right	Value right	RIP TOP RP
1	Rent	≥	300	≤	350	RIP
2						
...						
N						

"simplified" DPN – the complete freedom – is violated on purpose. All respondents have to evaluate the predefined set of attributes since some of the attributes are innovations.

The use of the simplified DPN as shown in Table 3.4 has a challenging feature: to evaluate experimental designs in a more rigid manner than is shown in Table 3.1. In the simplified DPN all respondents are confronted with the same set of shared attributes. Even the sequence of the attributes is under control. Hence, the distribution of X is known. Based on research we may postulate that the underlying abstract choice rule is also shared and therefore the choice of a theoretically sound probability distribution is clear. What differs is that the precise choice rules are still "hidden" in the DPN by means of the RIP, TOP, and RP. However, the hidden choice rules can be revealed if we systematically provide the DSS with all the profiles (p) of a full fractional design like in a conjoint approach. Instead of asking respondents the DPN responds. Likewise in the conjoint approach the output is a dataset of m acceptances and $n-m$ rejections for each profile out of the set p. This dataset can be regressed as we do in the algebraic models. As far as we know, no test has been performed so far to test whether this new feature is indeed justifiable.

To sum up, the added value of DPNs is twofold. They provide insight into the interdependency of preferences in the evaluation of supply. In addition, they show how people apply their preferences when there is a lack of supply. Despite the above advantages of the DPN, we nevertheless also found drawbacks. New applications of the simplified DPN have triggered new perspectives as well: the generalization problem and the role and explanation of meaning structures in housing choice.

Software Original DPN

The DPN management system (DPNMS) and the housing supply machine (WAM) were developed by Van Zwetselaar, Goetgeluk, and Floor. The tools and manual are freeware of the Faculty of Spatial Studies Utrecht University. It can be ordered via Dr R.W. Goetgeluk, ABF Research Delft, The Netherlands Roland.Goetgeluk@abf.nl, +31 15 2799349.

Acknowledgments In memory of my (co)promoters and colleagues: Frans Dieleman and Han Floor.

References

Arentze, T., & Timmermans, H. J. P. (2007). Parametric action decision trees: Incorporating continous attribute variables into rule-based models of discrete choice. *Transportation Research. Part B, 41*, 772–783.

Bettman, J. R. (1979). *An information processing theory of consumer choice*. Reading, MA: Addison-Wesley.

Boumeester, H., Coolen, H., Dol, K., Goetgeluk, R., Jansen, S., Mariën, G., & Molin, E. (2008). *Module Consumentengedrag WoON 2006*. Delft: Onderzoeksrapport OTB TU Delft in opdracht voor het Ministerie van VROM en de NEPROM.
Daalhuizen, F. B. C. (2004). Nieuwe bedrijven in oude boerderijen. De keuze voor een voormalige boederij als bedrijfslocatie (Dissertation, Uitgeverij Eburon, Utrecht).
Floor, H., van Kempen, R., & de Vocht, A. (1996). Leaving Randstad Holland: An analysis of housing preferences with DPNs. *Netherlands Journal of Housing and the Built Environment, 11*, 275–296.
Geuze, A., Goldhoorn, B., Hoornstra, A., & Middelkoop, G. W. (1994). Ontwerpend aan Holland; inspirerende schetsen en projecten voor het wonen in de 21ste eeuw. Almere/Amsterdam: Nationale Woningraad (NWR)/Stichting Architekten Onderzoek Wonen en Woonomgeving (STAWON).
Goetgeluk, R. W. (1997). Bomen over wonen. Woningmarktonderzoek met beslisingsbomen (Dissertation, Faculteit Ruimtelijke Wetenschappen Universiteit Utrecht, Utrecht).
Goetgeluk, R. W., & de Jong, T. (2005). Migration analysis as a politcal instrument: The case of the Leiden and the Bulb regions in Randstad-Holland. *Open House International, 30*(3), 75–81.
Goetgeluk, R. W., & Hooimeijer, P. (2002). The evaluation of DPNs for bRIPging the gap between the ideal dwelling and the accepted dwelling. Cybergeo, Systèmes, Modélisation, Géostatistiques, article 226, mis en ligne le 10 octobre 2002, modifié le 03 mai 2007.URL: http://www.cybergeo.eu/index2379.html.
Goetgeluk, R. W., Hooimeijer, P., & Oskamp, A. (1995). Modelling housing market search: empirical and algorithmic solutions. In M. M. M. Fischer, T. Sikos, & L. Bassa (Eds.), *Recent developments in spatial information, modelling and processing* (pp. 185–202). Budapest: Geomarket Co.
Heins, S. (2002). Rurale woonmilieus in stad en land.Plattelandsbeelden, vraag neer en aanbod van rurale woonmilieus (Dissertatie, Uitgeverij Eburon, Utrecht).
Heins, S., van Dam, F., & Goetgeluk, R. (2003). The psuedo-country-side as a compromised between spatial planning goals and consumer's preferences for rural living. *Built Envionment, 28*(4), 311–318.
Hooimeijer, P., & Linde, L. (1988). Vergrijzing, individualisering en de woningmarkt; het wodyn-simulatiemodel (Dissertatie, Faculteit der Ruimtelijke Wetenschappen, Rijksuniversiteit Utrecht).
Louwers, G. (1996). Het verplaatsingsgedrag van agrarische ondernemers (Doctoraal-scriptie, Landbouw Universiteit Wageningen/Landbouw-Economisch Instituut-DLO, Wageningen/Den Haag).
Mulder, C. H. (1993). Migration dymanics: a life course approach (PhD dissertation, Amsterdam University)
Mulder, C. H., & Hooimeijer, P. (1999). Residential relocation in the life course. In L. J. G. van Wissen & P. A. Dykstra (Eds.), *Population issues, an interdisciplinary focus* (pp. 159–186). Dordrecht: Kluwer/Plenum.
op 't Veld, D., Bijlsma, E., & Starmans, J. (1992). DPNs and expert systems tools, a new combination for application-oriented modelling of choice behaviour. *Netherlands Journal of Housing and the Built Environment, 7*(1), 101–124.
Oskamp, A. (1997). Local housing market simulation; a micro approach (Dissertation, Faculteit Ruimtelijke Wetenschappen, Universiteit Amsterdam, Nethur publicaties, Amsterdam).
Park, W., Hughes, C. R., Thukral, V., & Friedman, R. (1981). Consumers' decision plans and subsequent choice behavior. *Journal of Marketing, 45*, 33–47.
Timmermans, H., & van der Heijden, R. (1987). Uncovering spatial decision making processes: A decision net appraoch applied to recreational choice behaviour. *Tijdschrift voor Economische en Sociale Geografie, 78*(4), 276–304.
Timmermans, H. J. P., Molin, E. J. E., & van Noortwijk, L. E. (1994). Housing choice processes: Stated versus revealed modelling approaches. *Netherlands Journal of Housing and the Built Environment, 9*, 215–227.

van Kempen, R., Floor, H., & Dieleman, F. M. (1994). Wonen op maat; een onderzoek naar de voorkeuren en motieven van woonconsumenten en te verwachten ontwikkelingen daarin; deel 3: woonsituatie en woonwensen. Utrecht: Faculteit Ruimtelijke Wetenschappen, Universiteit Utrecht.

Wets, G. (1998). Decision tables in knowledge-based systems: Adding knowledge discovery and fuzzy concepts to the decision table formalism (PhD thesis, Eindhoven University of Technology).

Witlox, F. (1995). Qualitative housing choice modelling: DPNs versus decision tables. *Netherlands Journal of Housing and the Built Environment, 10*, 209–237.

Zwetselaar, M., & Goetgeluk, R. W. (1994). DPNs of housing choice: A critical evaluation of the reliability and validity of this technique. *Netherlands Journal of Housing and the Built Environment, 9*, 247–264.

Chapter 4
The Meaning Structure Method

Henny C.C.H. Coolen

4.1 Introduction

The meaning structure method is based on the means-end approach which was developed in the marketing discipline and which has mainly been applied for marketing and advertising strategy purposes (Reynolds and Olson 2001). This approach was introduced in housing research by Coolen and Hoekstra (2001) and it has been adapted and more fully developed for measuring the meaning of housing preferences by Coolen (2006, 2008). The purpose of the meaning structure method is to assess *what* people's housing preferences are and *why* they have these preferences. The approach, thus, explicitly uncovers both people's housing preferences as well as the motives for these preferences. In this respect the meaning structure method differs from most of the other approaches presented in this book, which focus mainly on *what* people want. The motives of people's housing preferences may be used for both marketing communication about houses as well as for developing and design purposes of dwellings by for instance real estate agents, designers, builders and developers.

In terms of the main dimensions for distinguishing between methods for measuring housing preferences the meaning structure method may be characterized as measuring stated preferences, the approach is non-mathematical and primarily concerned with housing attributes. The outcome is formed by a set of preferred housing attributes, a set of motives for preferring these attributes, and a set of relations between the attributes and the motives.

In this chapter, the original means-end approach is described first. It forms the foundation of the meaning structure method and is still used in housing research (Lundgren and Lic 2010). Subsequently the meaning structure method, based on several adaptations to the means-end approach, is presented.

H.C.C.H. Coolen (✉)
OTB Research Institute for the Built Environment,
Delft University of Technology, Delft, The Netherlands
e-mail: h.c.c.h.coolen@tudelft.nl

S.J.T. Jansen et al. (eds.), *The Measurement and Analysis of Housing Preference and Choice*, DOI 10.1007/978-90-481-8894-9_4,

4.2 Original Means-End Approach

The means-end approach consists of a conceptual model, a measurement procedure named laddering, and an analysis method. Each of these aspects will be described in turn. The examples in this section come from the pilot study by Coolen and Hoekstra (2001).

4.2.1 Conceptual Model

The focus of the means-end approach is on the relationships between goods and consumers. A good is defined by a collection of attributes, which in the case of a dwelling may be such attributes as the number of rooms, the size of the living room, and the presence/absence of a garden. These attributes yield consequences when the good is used. The importance of the consequences is based on their ability to satisfy the personally motivating values and goals of people. In this context, values are defined as "desirable transsituational goals, varying in importance, that serve as guiding principles in the life of a person or other social entity" (Schwartz 1994).

A means-end chain (MEC) is a model that provides a way of relating the preference for a good to its contribution to the realization of objectives and values. Means in this context are goods that people consume and activities that they carry out. Ends are positively evaluated situations such as freedom, privacy and friendship. The most important linkages between values and objectives on the one hand and preferences on the other form the elements of the means-end chain model. The original means-end chain model is based on four assumptions (Gutman 1982).

The first assumption states that objectives and values influence preference and choice processes. This means that people's behavior is considered as goal-directed and value-oriented. Secondly, it is assumed that people can keep track of the enormous diversity of goods by grouping them in sets or classes so as to reduce the complexities of choice. This means that consumers cannot only classify goods into product fields (housing, cars, holidays, for example), but are also capable of creating functional classifications. An example of such a functional classification is 'preserving my image', that might contain the objects 'detached house', 'Jaguar' and 'luxury cruise'. Third, it is assumed that the behavior and actions of consumers have consequences, although these consequences do not have to be the same for everybody. For instance, a consequence of a house with five rooms for one household might be that every household member has their own room, and for another household the consequence may be that there is a guestroom and a study. Finally, there is the assumption that consumers learn to associate particular consequences with particular behaviors.

In the original model the term *consequences* is used where we may also speak about *goals* or *objectives*. The terms consequences, goals and objectives will be used interchangeably in this chapter. By *consequence* we mean every direct or indirect result of a person's behavior. Consequences can be desirable or undesirable. Desirable consequences are also known as *benefits*. The central idea in the means-end model is that consumers choose the actions which are expected to produce the desired consequences and which minimize the undesirable consequences. Values provide consequences with a positive or negative valence. Therefore, the linkage between values and consequences is of essential importance in the means-end chain model. A certain good must be consumed to realize a desirable consequence, but in order to do that a choice must be made among alternative goods. To be able to make this choice, the consumer must learn which goods possess the attributes that produce the desirable consequences. Thus, the second essential linkage in the model is the one between consequences and the attributes of goods.

The original and simplest means-end chain model has three levels: product attributes – consequences – values. A simple example of a means-end chain model related to housing would be: five rooms (attribute) – more space (consequence) – privacy (value) (see Fig. 4.1).

Although means-end chains with more than three levels have been described in the literature (Walker et al. 1987), we restrict ourselves to Gutman's original model (1982) with three levels.

In the context of means-end theory, the categorization process is considered to be the manner in which consumers organize their thinking about specific goods. It is assumed that consumers create classes of goods that are instrumental in bringing about certain consequences and that contribute in their turn to the achievement of

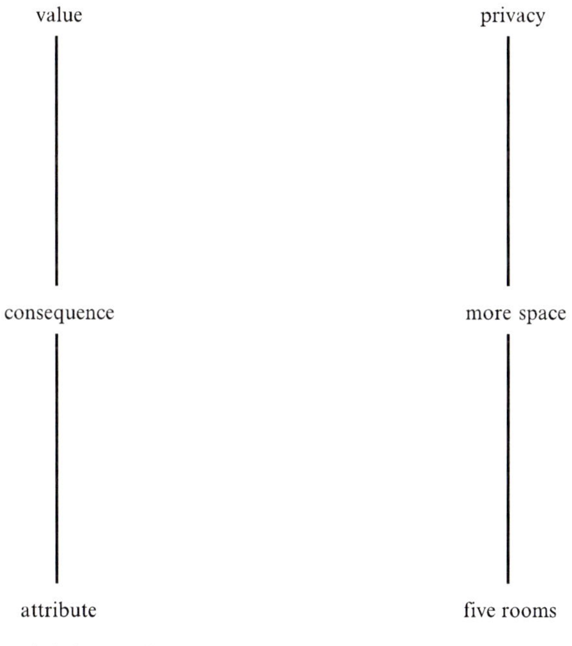

Fig. 4.1 Means-end chain model

valued end situations. The categorization process forms the way in which people segment their complex environment into meaningful classes (Rosch 1978). Through categorization people divide their environment into smaller units that they can deal with more easily. This categorization process is necessary, because the environment comprises many more objects than people have values. For consumers it is essential to reduce the complexity in the multiplicity of goods that the market offers to avoid information overload and enable further processing. The categorization process takes place at every level of the means-end chain and results in terms of abstractness in a hierarchy of categories. At lower levels of the chain the categories are less abstract, because they consist of goods that are similar with respect to more concrete attributes. At higher levels of the chain the categories are more abstract because they are more inclusive, which means that they contain different products that are only similar with respect to a more abstract attribute. If consumers distinguish between detached and non-detached dwellings two relatively concrete categories of lower inclusiveness would be created. If detached dwellings were grouped with Jaguar and luxury cruise into a category 'status' a more inclusive category would be created. If the achievement of values is sought, classes of products must be systematically related to higher objectives, because otherwise there can be no question of instrumentality. Although the division into classes is based on the attributes of goods, the choice of the attributes that are important for a consumer is determined by his or her values. Goods are thus divided into various classes on the basis of both the attributes that are emphasized and the attributes that are ignored. The manner in which consumers identify or describe goods therefore fits with their classification of these goods into functional classes. Abstract values that come high in the means-end chain have to be translated through less abstract objectives to consequences and attributes, thus providing the basis for the creation of classes of goods. Consumers therefore create categories and classifications of goods so that they can contribute as much as possible to the realization of desired consequences and the attainment of values.

Following Schwartz' (1994) conceptualization values are conceived as objectives, which, consciously or unconsciously, function as criteria in all our actions. They have cognitive, affective and behavioral aspects (Rokeach 1973). In this notion of values as objectives we can recognize the following aspects: (1) values function as interests for individuals or groups; (2) values motivate behavior and give it direction and intensity; (3) values function as criteria for the evaluation and justification of behavior; (4) values are acquired through the socialization of dominant group norms and through unique individual experiences (Schwartz 1994).

In order to be able to live and function in a social environment, individuals and groups transform the needs, inherent to human existence, into specific values. From these fundamental human needs, Schwartz (1992, 1994) derives ten universal, motivational value domains. These domains, with, in brackets, some values belonging to each are:

1. Power (social power, wealth);
2. Achievement (successful, ambitious);
3. Hedonism (pleasure, enjoying life);

4. Stimulation (daring, exciting life);
5. Self-direction (independent, curious);
6. Universalism (social justice, unity with nature);
7. Benevolence (helpful, true friendship);
8. Tradition (humble, devout);
9. Conformity (politeness, self-discipline);
10. Security (family security, clean).

Every individual strives for values belonging to each of these domains. According to Rokeach (1973), the values will not be of the same importance for every individual. In other words, individuals organize and structure their values so that they are in a position to choose from alternative objectives and actions and are able to resolve potential choice conflicts. Such a configuration of values is called a *value system* (Rokeach 1973). Value systems are relatively stable in the sense that over a longer period of time they will on average comprise the same values. Changes in value systems do not occur so much in the values of which they are comprised as in the relative importance ascribed to every value within the system (Rokeach 1973).

It is generally acknowledged (Rokeach 1973; Williams 1979; Bettman 1979; Schwartz 1996) that values can influence behavior in various ways. For example, values contribute to our ability to take a standpoint with respect to political and social questions. They may be used in the assessment of ourselves and others. Furthermore, values play a central part in comparison processes, and they may form criteria for the evaluation of the opinions, attitudes and actions of ourselves and others. In a choice situation, various values will be activated in a person's value system. However, it is unlikely that people will be able to act in agreement with all of the activated values simultaneously. In this context, a value system is a learnt and organized entity of principles and rules that helps people, in choosing between alternatives, to resolve conflicts with respect to a choice and to take decisions. A value system is thus a cognitive system of which only a relevant part becomes activated.

Given the means-end chain model, as discussed in this section, the measurement of means-end chains will be described next.

4.2.2 Measurement Procedure

The measurement of means-end chains takes place in five phases:

1. Elicitation of the attributes;
2. Selection of the attributes;
3. Elicitation of the attribute levels;
4. Performing laddering interviews;
5. Determination and coding of means-end chains.

These phases are discussed below. The data used to illustrate several aspects of the measurement and analysis process come from a pilot project (Coolen and

Hoekstra 2001). The purpose of this pilot was to investigate the feasibility of the means-end approach for research on housing preferences. For the pilot project ten respondents, who were considered knowledgeable with respect to housing, were interviewed. All interviews took place in the home of the respondent by one of the researchers.

4.2.2.1 Elicitation and Selection of Attributes and Attribute Levels

The first phase in measuring means-end chains concerns the elicitation of relevant attributes for the laddering interview. Often the Repertory or Kelly Grid (Kelly 1955) is used for this. In this procedure the respondents are presented with a limited number of triads with constantly differing products from a particular product field. For every triad they must indicate in what way two of the three named products are similar to each other and consequently in what way they differ from the third product. By repeating this procedure for every triad, it becomes clear what the salient characteristics of the product field are for the respondent and what the poles of each of these characteristics are. For instance, a respondent presented with three dwellings may indicate that two of the dwellings are alike because they are detached houses, and that they differ from the third dwelling because that is a non-detached house. Here the salient characteristic is the level of detachedness of a dwelling and the poles are detached and non-detached. This method is often used when the relevant attributes are unknown. In addition, the method can be readily implemented if one is dealing with a relatively homogeneous product field and/or if a product field consists of readily recognizable brands, because the procedure will then most likely result in a relatively small number of salient features. However, this does not apply to the product field of housing. A house is an extremely heterogeneous product and brands are hardly known. Moreover, much is known about relevant housing attributes (Clark and Dieleman 1996). That is why we worked with lists of housing attributes ourselves, based on the relevant literature and the goal of the research.

The second phase comprises the selection of attributes. The respondents were assigned the task of selecting from the list of attributes those that were most important for them. In addition, they had the possibility to mention attributes they considered important that were not on the list. No limit was set to the number of attributes that could be chosen. If a respondent chose more than eight attributes, he/she was then assigned the task of selecting the eight most important ones. This was done because otherwise the interview burden from the laddering interviews might become too great for the respondents.

In the third phase, the respondents were asked which level of each of the selected attributes they preferred. If for example the number of rooms was a selected attribute, then the respondent was asked how many rooms he/she would like. The preferred level, which serves as the starting point for a laddering interview, was determined for every selected attribute.

4.2.2.2 Laddering Interviews

The key phase in measuring means-end chains is the fourth. In this phase the actual means-end chains are determined. For this purpose, a semi-structured interviewing technique known as laddering is used. It involves a tailored interviewing format using primarily a series of directed probes, typified by the '*Why* is that important to you?' question, with the express goal of determining the links between the essential elements of a means-end chain: attributes – consequences – values. A respondent who states that he/she wants a house with five rooms would then be asked: 'Why do you find it important that the house you want should have five rooms?'. The *why* question is repeated as a reaction to the answer of the respondent.

The process stops when the respondent can no longer give any more answers to these *why* questions. Letting the interview begin at the concrete level of the attributes and then continuously asking *why* allows the underlying consequences and values of a certain choice to be brought into the open. In this way, a means-end chain can be determined for each respondent and each attribute level; such a chain is called a *ladder*. A ladder shows the underlying reasons of the preference for a certain attribute level. This yields insights into the classifications employed at higher levels of abstraction and may reveal how the properties of goods are processed from a motivational perspective.

Since the respondents are asked to be introspective and to talk about their motivations, a non-threatening interview environment must be created. This can be facilitated during the introduction to the interview by pointing out to a respondent that in the context of this type of research there is no such thing as a correct or incorrect answer. It is primarily the respondent's opinion that is important. Thus, the respondent is positioned as an expert and the interviewer fulfils the role of a *facilitator*, who has to keep the respondent talking. Furthermore, it is of great importance that the interviewer is able to identify the relevant elements of the respondent's answers. This means that the interviewer needs to be fully acquainted with the means-end chain model and the content matter to which the interview refers.

The ten laddering interviews we performed in the pilot study were recorded on tape and subsequently transcribed. The researchers performed most of the transcriptions themselves. In the case that someone else performed a transcription, one of the researchers checked it thoroughly. During our interviews, respondents quite frequently gave so-called *forked answers* (Grunert and Grunert 1995). This means that several consequences are linked to only one attribute. According to Grunert and Grunert (1995), this occurs most often with respondents who have thought thoroughly about a certain preference or decision and consequently have an extensive meaning structure in the area concerned. This is almost certainly the case for our knowledgeable respondents. However, the high incidence of 'forked answers' in our pilot project might also be specifically related to the product field of housing. After all, a house is a good in which the consumer is seriously involved, which makes preferences and decisions in this area mostly well thought through. If respondents gave a forked answer, efforts were made to determine a separate ladder for every named consequence.

4.2.2.3 Constructing Means-End Chains: from Interviews to Ladders

In the fifth phase, the means-end chains are determined on the basis of the interviews. The raw data generated by the laddering interviews are the (transcribed) verbalizations of the respondents. First, a content analysis was carried out on these free responses. This resulted in a set of ladders for each respondent. Subsequently, the elements of these means-end chains were coded, dividing them according to topic and level in the hierarchy (attribute, consequence, value). In this process, several choices about the interpretation of the various elements of the ladders had to be made. To reach as much intersubjectivity as possible, several researchers were involved in the construction of the ladders from the interviews and the subsequent coding of these ladders. Four researchers constructed and coded the ladders of the first four interviews. After that, the ladders each researcher had constructed and coded were compared with each other in two sessions in which all four researchers participated. Possible differences were discussed until agreement was reached. Furthermore, this consultation process resulted in a coding scheme for the remaining six interviews. For these interviews, ladders were first constructed and coded by two researchers separately. Subsequently, the results were compared with each other and differences were resolved. For the coding of the values that appeared in the laddering interviews, the value domains and values of Schwartz (1992, 1994) were used as a frame of reference. All the values found fitted into this framework.

Some examples of ladders that were derived from the interviews are shown in Fig. 4.2. In this figure, all the means-end chains start at the level of attributes and end at the level of values. However, this does not necessarily have to be the case (see also Fig. 4.3). Sometimes the value level is not reached and the chain stops at the level of consequences. There may be several reasons for this. Firstly, it is

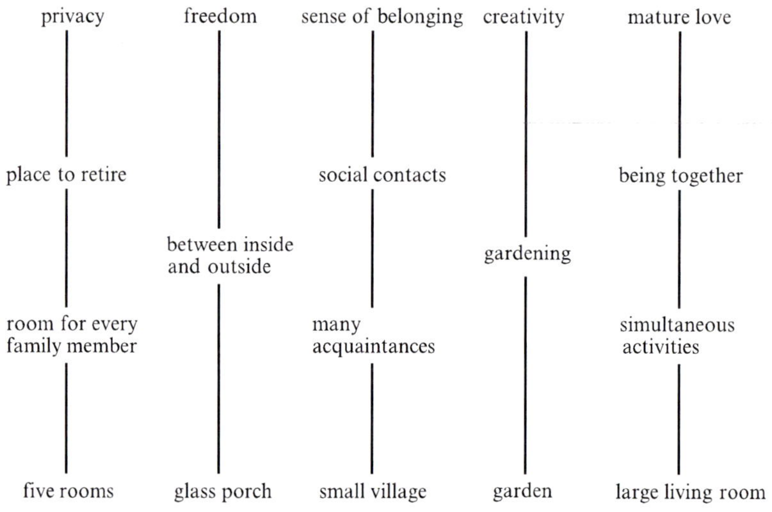

Fig. 4.2 Examples of ladders

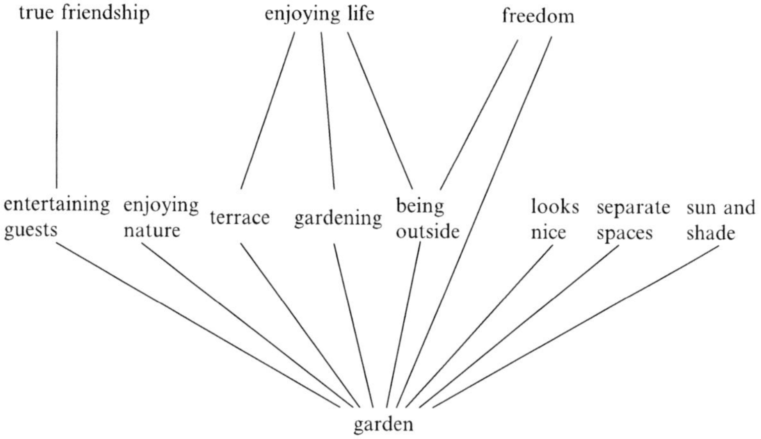

Fig. 4.3 Hierarchical value map of the preference for a garden

possible that the interviewees got stuck at the level of consequences. Secondly, the interviewers may not have pursued the questioning deeply enough, which, considering their unfamiliarity with the laddering method, is not inconceivable. Thirdly, it may be the case that the respondent's motivations for a certain attribute are only formed by consequences.

The most remarkable thing about the ladders in Fig. 4.2, though, is the fact that the consequences that are mentioned differ tremendously. Some are functional (a room for every family member), while others are more psychosocial in nature (place to retire). Several consequences are rather concrete (gardening, room for every family member) and others more abstract (social contacts). We also note that some of the consequences concern everyday activities (gardening), while many others (between inside and outside, a room for every family member) do not.

4.2.3 *Analysis Method*

4.2.3.1 Construction of a Hierarchical Value Map

The ladders of the individual respondents are represented together in a hierarchical value map, which is a kind of a tree diagram. In order to be able to construct such a hierarchical value map the ladders of the individual respondents are aggregated by means of an implication matrix. An implication matrix is a square matrix that represents the relationships between the elements from the ladders. The rows and the columns of the matrix are formed by the elements from the ladders arranged into attributes, consequences, and values. The cells of the implication matrix show the number of direct links between the elements of the ladders. In the literature, examples of implication matrices and hierarchical value maps are not abundant.

One of the few examples we know of can be found in Reynolds and Gutman (1988), where means-end theory is applied to the product field of wine-coolers. Their analysis resulted in a 23×23 implication matrix and a well-organized hierarchical value map. A preliminary analysis of the results of our ten interviews, however, revealed more than 50 attributes and/or attribute levels and approximately 150 different consequences.

The variation in attributes, consequences and values depends, of course, on the level of detail at which the coding process is stopped. If this process ends at a relatively detailed level, the loss of information will be limited and the resulting number of categories will be large. As a consequence of this, the implication matrix will be large, the cell frequencies will be relatively low, and the construction of a hierarchical value map by means of a paper-and-pencil method will be complicated, if not impossible. In such a case, it remains doubtful whether computer-aided means could ease the burden, because the resulting hierarchical value maps are likely to be very complex and thus difficult to interpret.

Of course, the size of the implication matrix could be reduced by using broader coding categories. For our pilot project we did not consider this an appropriate option. The variation in attributes and consequences was so big that a less detailed coding would have resulted in an in our view unacceptable loss of information and a very general, and therefore possibly less meaningful, hierarchical value map. That is why we chose another solution; we decided to construct implication matrices and hierarchical value maps for separate attributes. As an example, the implication matrix of the feature garden is presented in Table 4.1. The rows and columns represent the attribute garden, and the consequences and values that are associated with this attribute. The numbers in the cells indicate the number of times a direct relation between the appropriate row and column element was observed in the ladders. For instance, the link between garden and being outside occurs in ten ladders, and the link between being outside and enjoying life has been observed seven times.

The dominant connections in an implication matrix can be represented graphically in a tree diagram known as a *hierarchical value map (HVM)*. To construct such a tree diagram Reynolds and Gutman (1988) describe a paper-and-pencil method, which we also applied in the pilot study. In this context a distinction is made between ladders and chains; the term ladders refers to elicitations from individual respondents, the term chains is used in reference to sequences of elements that emerge from the implication matrix. The construction of a HVM from the implication matrix begins by considering adjacent relations, that is, if A and B, and B and C, and C and D are linked, then a chain A – B – C – D is formed. Notice that there does not necessarily have to be an individual with an A – B – C – D ladder for an A – B – C – D chain to emerge. A HVM is gradually built up by connecting all the chains that are formed by considering the linkages in the implication matrix. The most typical approach is to map all the relations above a certain cutoff level, the size of which depends on the size of the sample. To actually construct a HVM one must literally build up the map from the chains that are extracted from the implication matrix. In constructing the HVM in Fig. 4.3 from the data in Table 4.1 the most efficient way is to start in the first row, for which there is a value above the

Table 4.1 Implication matrix of the feature garden

	1.	2.	3.	4.	5.	6.	7.	8.	9.	10.	11.	12.	13.	14.	15.	16.	17.	18.	19.	20.
1. Garden		5	10	2	3	9	3	1		1	4	2	3							
2. Terrace			1						1		1			2						
3. Being outside							1		1				2	7						
4. Looks nice						1								1	1					
5. Separate spaces																			1	
6. Gardening				1			1			1				4	1					1
7. Enjoying nature														1						
8. More space																		1		
9. Cosiness																				
10. No discomfort													1	1						
11. Entertaining guests									1	1						1	2			
12. Sun and shade			1											1						
13. Freedom																				
14. Enjoying life													1							1
15. Clean																				
16. Belonging																				
17. True friendship																				
18. Choose own goals																				
19. Benevolent																				
20. Creativity																				

chosen cutoff level (cutoff = 1 in our example), and choose the first entry higher than the cutoff level. In Table 4.1 this is the link between garden and terrace with a value of 5. Next, one moves to the second row (terrace) to find the first entry exceeding the cutoff level. In Table 4.1 this is the link between terrace and enjoying life. Thus, the chain has now become garden – terrace – enjoying life. Continuing in the same manner in Table 4.1 we find that the chain will not extend anymore. Having reached the end of the chain, one goes back to the beginning to see if there are other salient relations in the same rows of the matrix that have already been inspected. For example, inspecting the first row we find that garden is also related to being outside. Next, we move to the third row (being outside) to find that being outside is related to freedom. Here the chain garden – being outside – freedom ends. The next step is to move back to the first row and continue with the process. After finishing the first row, we move to the second row and start the process over again. Subsequently, the process is repeated in all the rows of the implication matrix. The result of doing this for the implication matrix in Table 4.1 is the hierarchical value map in Fig. 4.3, which represents all the salient relations in the implication matrix.

Figure 4.3 shows the hierarchical value map of the attribute *garden*, which is based on a 20 × 20 implication matrix. We chose this attribute because almost all the interviewees (nine of the ten respondents) mentioned it. The hierarchical value map clearly illustrates the great variation in consequences and values we encountered in our pilot project. The attribute 'garden' is linked with one other attribute (terrace), seven consequences (separate spaces, gardening, looks nice, being outside, entertaining guests, sun and shade and enjoying nature) and three values (enjoying life, freedom, true friendship). This implies that there are not necessarily one-to-one relations between the different elements of the hierarchical value map. As we can see in Fig. 4.3, different consequences may contribute to the accomplishment of one and the same value, and one consequence may also contribute to the realization of different values. This is as one might have expected on the basis of Rokeach's (1973) ideas about values and value systems, which have been summarized in Sect. 4.2.1.

In analyzing several of the individual ladders the most remarkable thing that attracted our attention was the tremendous variation in the type of consequences that appeared in the ladders. After aggregating the individual ladders for the attribute garden this phenomenon does not disappear. The consequences that appear in the hierarchical value map in Fig. 4.3 concern everyday activities (gardening, being outside), functional consequences (separate spaces, sun and shade), psychosocial consequences (looks nice, enjoy nature) and even another attribute (terrace). This seems to support several aspects of means-end theory and our earlier conclusion, based on the analysis of the individual ladders, that factors that intervene in the relationships between housing attributes and values comprise a diversity of different types of consequences.

We can also see in Fig. 4.3 that the attribute 'garden' is related, mainly indirectly, to a variety of values. These values are indicators of different value domains. In terms of Schwartz's (1992, 1994) conceptualization, the attribute 'garden' is related to the value domains Self-direction (freedom), Hedonism (enjoying life) and Benevolence (true friendship). From a motivational point of view this means that both inner-directed (hedonism, self-direction) and outer-directed drives (benevolence) seem to motivate the preference for a garden.

Both the values as well as the consequences may be used for marketing communication purposes (Vyncke 2002), for instance by real estate agents, builders and developers, and garden centers in the case of the garden. Values such as freedom, enjoying life and true friendship are important values in Dutch society that appeal to many people and with which many people also want to be associated. The same can also be said about many of the consequences that appear in Fig. 4.3. Consequences such as entertaining guests, enjoying nature, gardening, and being outside are activities that many people can identify with.

From the point of view of developing and designing gardens, the great variety of consequences seems to indicate that standard gardens do not exist. Apparently, people want their garden to serve many diverse functions and combinations of functions. Although this variation in functionality is evident from Fig. 4.3, it is also clear that several of the consequences are not specific enough for design purposes. For instance, terrace, separate spaces, and sun and shade give an indication of what people want in their garden, but, from a designer's perspective, can only be starting points that need further elaboration.

An interesting aspect of Fig. 4.3 is the direct relationship between the attribute 'garden' and the value 'freedom'. Apparently, the association between having a garden and feeling free is so strong for our respondents that there are no intervening consequences that emerge immediately from the hierarchical value map. This empirical result seems to undermine certain aspects of the means-end approach. We shall come back to this in the next section.

4.3 The Meaning Structure Method

The meaning structure method contains important adaptations to the means-end approach. These adaptations result from both our experience with using the original approach in housing research and from methodological considerations with respect to the analysis of means-end chains. In this section the major adaptations to the conceptual model, measurement procedure and analysis method are described.

The data used to illustrate the meaning structure method were collected for a project with the aim of comparing the meaning structures of residential environment preferences of urban and suburban apartment dwellers. For this purpose, two geographically dispersed locations were selected. The suburban area chosen was a disused airport on the outskirts of The Hague, a large area where construction is still going on. The urban area selected is located in the city of Rotterdam. It was constructed in the middle of the nineties as part of a master plan for the development of former harbor districts.

A total of 45 semi-structured interviews, all of which were taped, were conducted at the respondents' homes. For the purpose of this paper, it is not necessary to distinguish between the two sub-populations, so the dataset is treated as one.

4.3.1 Conceptual Model

In the pilot study by Coolen and Hoekstra (2001) it was found that the original MEC-model does not always represent the relations between housing features and dwellers well. For instance, sometimes there are no intervening consequences, which makes the relationship between attribute and value direct. A good example is the direct relationship between the attribute garden and the value freedom in Fig. 4.3. And sometimes there seem to be no values involved, leaving only a relation between attribute and consequence, for instance the relationship between garden and separate spaces in Fig. 4.3. It seems that the MEC-model needs to be adapted in order to also make these relationships possible. According to Coolen (2008) the need for adjustment of the means-end model is, at least partly, due to the values category, which is a rather narrowly defined category making everything not being a value into a consequence. A less narrowly defined scheme of categories would make the means-end model more flexible. Rapoport (1988) provides such a scheme. His focus is on the relationship between people and the environment, which is a more inclusive relationship than the one between consumers and goods, and he emphasizes the importance of meaning in understanding the built environment. Meaning is one of the central mechanisms in linking environments and people by providing much of the rationale for the ways in which environments are *shaped* and *used*. He also argues that the common distinction between function and meaning is misguided, because function has mainly been identified with manifest aspects of the environment, while more latent aspects may also help us understand built form. This implies that meaning is not only part of function, but is often the most important function of the built environment. This functional approach to meaning makes Rapoport's conceptualization very similar to the means-end approach. In the means-end approach goods are functional for the achievement of consequences and values, and in Rapoport's conceptualization environmental objects are functional in terms of the meaning they have for people. Rapoport (1988) distinguishes three levels of meaning in the built environment. *High-level* meanings are related to cosmologies, world views, philosophical systems, etc., similar to Schwartz's value domains; *middle-level* meanings such as identity, status, wealth, power, etc. which are also called latent functions, and which concern the latent aspects of activities and behavior; *lower-level,* everyday and instrumental meanings, for example accessibility, privacy, seating arrangements, movement, etc. which are also called manifest functions. The distinction between manifest and latent functions is similar to the distinction between consequences and values, but the categories of manifest and latent functions are more general and inclusive categories. These categories are consequently used in the meaning structure method, which is presented in Fig. 4.4, and the chain attribute – manifest function – latent function is called a meaning structure (Coolen 2006).

The great advantage of using the more analytical categories of manifest and latent functions instead of the categories of consequences and values is that what might be a manifest function for one individual, for instance freedom, may be a

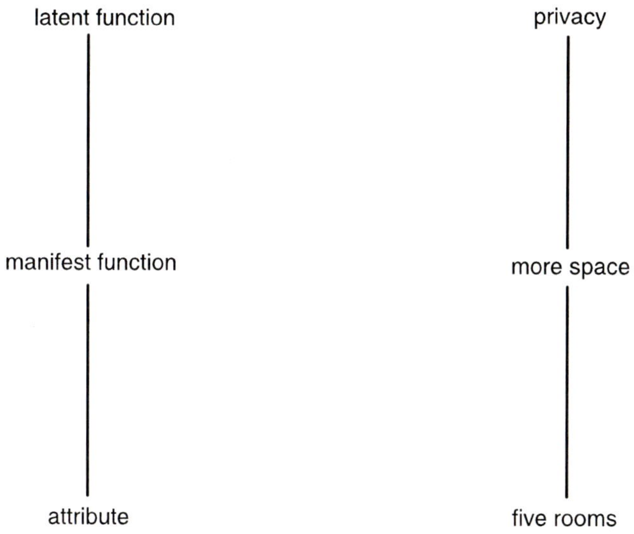

Fig. 4.4 Meaning structure model

latent function for another individual. This makes the meaning structure method an approach that truly considers the functions of attributes from the perspective of the individual.

4.3.2 *Measurement Procedure*

The measurement procedure for measuring the meaning structures of preferences for housing attributes is a partly adapted version of the procedure for the determination of means-end chains, which has been described in Sect. 4.2.2 above. Although all the steps described there could be passed through, the measurement of the meaning structures of housing preferences generally takes place in three phases:

1. Elicitation of the salient housing attributes;
2. Elicitation of the (preferred) levels of the salient housing attributes;
3. Measurement of the meaning structures.

4.3.2.1 Elicitation of Salient Attributes and Attribute Levels

The first step in measuring the meaning structures concerns the elicitation of salient housing attributes. Many elicitation methods are available that range from letting the respondents mention the features themselves, to presenting the respondents with a list of features (cf. Reynolds et al. 2001). Since much is known about important dwelling features we often use lists or sets of cards with relevant housing features. Respondents have to select the most important features from these sets. They also have the possibility to add features they consider important and that are not on the list or cards, enabling them to determine exactly what a dwelling is to them. The choice to use lists or cards with features was supported by the very large number of dwelling features. It was expected that, because of the limited information processing capability of human beings, list or sets of cards would support the respondents in conceptualizing their important dwelling features.

In the second phase the respondents are asked to indicate which level of each of the salient features they prefer. Sometimes it may be more fruitful to let respondents indicate what they do not prefer. For instance, when the attribute type of dwelling is a salient feature, either the preferred type may be indicated or the dwelling type that is definitely not wanted. Allowing respondents to indicate what they definitely do not prefer, their so-called non-preference, is particularly relevant for situations in which the respondent cannot articulate their preference for a certain level of a salient feature very well. For example, some respondents know very well that they do not want to live in an apartment, but have no clear preference for either a dwelling in a row or a semi-detached dwelling.

4.3.2.2 Laddering Interviews

The starting point for determining the meaning structure of each salient dwelling feature is the preferred or non-preferred level of that feature. The meaning structures are measured, in the third phase, by the semi-structured interviewing technique known as laddering (Reynolds and Gutman 1988). The interview proceeds according to the tailored format described above using primarily a series of directed probes of the form 'Why is that important to you?' The purpose of this interviewing format is to determine the relationships between on the one hand the preferred or non-preferred level of a salient feature and on the other hand the manifest and latent functions this housing feature has for the respondent. Letting the interview begin at the preferred or non-preferred level of a salient housing feature and subsequently proceeding with several why questions allows the most closely associated meanings of the feature to be revealed. In this way meaning structures can be determined for each salient housing feature level and for every respondent. In contrast with the original laddering interviews the meaning structures are now constructed during the interview by the interviewer and the respondent together on paper. There are good reasons for constructing the meaning structures in this way. Writing each answer down on paper gives the respondent some time during the interview to reflect about his or her answer and to explore and discover other aspects of the cognitive structure under construction. It also gives the interviewer some time to reflect about the answer and to make sure he/she understood the answer correctly. If necessary, the interviewer can probe the respondent about the exact meaning of his answer. Furthermore, instead of being an interviewee who only has to answer questions passively, the respondent has a more active role in the interview and this involvement may work as a motivating factor. Moreover, the time needed for processing the data is diminished considerably by having the meaning structures on paper immediately after the interview is finished.

4.3.2.3 Processing the Interviews

The raw data generated by the laddering interviews, both on paper and tape, are the verbalizations of the respondents. These verbalizations are so-called less-structured data, which can only be used for further analysis – description, interpretation, explanation, mathematical and statistical analysis – when they contain some minimum level of structure called a category system (Coolen 2007). The process of developing such a category system is called categorization, which is carried out on the raw data by means of content analysis. The content analysis is performed on the meaning structures on paper, while the taped interviews are only consulted if there are doubts about the interpretation of a meaning structure that is on paper. The content analysis results in a set of meaningful categories that will be used for all respondents in the subsequent analysis. Subsequently, the meaning structures of each respondent are coded according to the set of categories. In this process, choices about the interpretation of the various elements of the meaning structures often have to be made. To reach as much intersubjectivity as possible, two researchers were involved in the construction of the categories from the interviews and the subsequent coding of the meaning structures. The categories and meaning structures each researcher has constructed and coded are compared with each other and possible differences are discussed until agreement is reached. The meanings that the respondents associate with the housing features dwelling type apartment, number of rooms, and size of living room and the frequency with which these meanings were mentioned are represented in Table 4.2.

Table 4.2 The meanings of three dwelling features

	Dwelling type = apartment (n = 28)	Size of living room (n = 32)	Number of rooms (n = 32)	Total
Freedom	11	14		25
Privacy	6	3	14	23
Comfort	25	12	14	51
Space	15	21	13	49
Enjoying life	9			9
Social contacts	6	9	13	28
Health	11			11
No garden	8			8
Furnishings	7	21	14	42
Multi-functionality		5	24	29
Well-being	9	23	9	41
Clean			7	7
Tradition			1	1
Family			9	9
Total	107	108	118	333

4.3.3 Analysis Methods

Given the categorization and coding of the meanings the originally less-structured data can be analyzed in much the same way as structured data, because categorization is a form of nominal measurement (Coolen 2007). The category systems can be displayed in two general formats, namely matrices and networks (Miles and Huberman 1994), and for the analysis of both types of displays essentially the same data analysis techniques can be used as with structured data (Handwerker and Borgatti 1998; Ryan and Bernard 2000). So, two types of data analysis may be performed on the data: symmetrical and asymmetrical analyses. In a symmetrical analysis the structural aspects of the data, i.e. the links between the meanings, are ignored. If the purpose of the analysis is, for instance, to find similarities and differences between the meanings of a feature or to find similarities and differences between subgroups of respondents with respect to the meanings of a feature a symmetrical analysis is the appropriate way to proceed. Table 4.2 is an example of a simple symmetric analysis.

In an asymmetrical analysis of the coded meanings one takes the structural relations between the meanings explicitly into account, as is the case in a hierarchical value map, and by doing so one can construct two types of representations with meaning structures. One type represents only individual meaning structures (see Fig. 4.2), the other type contains the relationships between the meaning structures of all respondents and is called a meaning network. Traditionally this last type of relationship is represented in a hierarchical value map, which was originally constructed by means of the paper-and-pencil method that was described in Sect. 4.2.3, while computerized methods became available later on. Valette-Florence and Rapacchi (1991) called the paper-and-pencil procedure cumbersome and boring, and, because it is conducted in a stepwise manner it relies on a trial-and-error basis, which can lead to various omissions and errors. In order to avoid these drawbacks they propose taking full advantage of graph theory techniques, and they develop a method called weighted acyclic network analysis, which aims to construct a hierarchical value map by graph-algorithmic means. This method is not described here, because more general graph theoretic methods were used. By recognizing that an implication matrix is very similar to an adjacency matrix Coolen (2006) opened up the full graph analytic toolbox for application to meaning structures. One of the consequences of doing this is that the hierarchy assumption, which assumes a hierarchical ordering of the different categories in terms of abstractness, is dropped, and that the presence or absence of hierarchy in a meaning network becomes an empirical matter. Van Rekom and Wierenga (2007) have investigated this hierarchy assumption in means-end theory, and for the cases they studied the hierarchy assumption had to be rejected. Moreover, as Cohen and Warlop (2001) have argued, in many research situations one does not have to assume hierarchy in means-end relations at all in order to achieve meaningful and relevant answers to research questions.

The individual meaning structures, which are relational data, form the basis for the construction of a meaning network. Since some structural aspects of meaning networks are discussed in this paper, some terminology about networks is outlined next (cf. Wasserman and Faust 1994). A meaning network is constructed from the individual meaning structures by means of a so-called adjacency matrix. An adjacency matrix is a square matrix that represents the relationships between the meaning categories from the meaning structures. The rows and columns of the matrix are formed by the meanings, and the cells of the adjacency matrix show the number of direct links between the meanings in the individual meaning structures. The adjacency matrix of the dwelling feature number of rooms is shown in Table 4.3, in which we see, for example, that eight respondents have mentioned the link between multi-functionality and privacy.

The connections between the meaning categories can be represented graphically in a valued digraph – a network representation – in which the meanings are represented as nodes n_i and the directed links between them as arcs l_k. Associated with each arc is a value v_k that indicates the number of times that the link between the two nodes connected by the arc has been observed. The value in cell (i, j) of Table 4.3 represents the number of observed arcs directed from the meaning in row i to the meaning in column j. The graphical display of the relations in Table 4.3 is called a meaning network, and the meaning network for the dwelling feature number of rooms is depicted in Fig. 4.5; the thicker the link between two meanings in this figure, the more often the relation between those meanings is mentioned by the interviewees. Clearly, the most mentioned link is the one between multi-functionality and privacy, which is confirmed by Table 4.3.

The indegree $d_I(n_i)$ of a node n_i is the number of nodes that are adjacent *to* n_i, so indegree is the number of arcs terminating at n_i. The indegree of a particular meaning is the number of times that the meaning is the destination of a connection with other meanings. Indegree is the column sum of a meaning in the adjacency matrix. The outdegree $d_O(n_i)$ of a node n_i is the number of nodes that are adjacent *from* n_i. The outdegree is thus the number of arcs originating from n_i. The outdegree of a particular meaning is the number of times the meaning is the origin of a connection with other meanings. Outdegree is the row sum of a meaning in the adjacency matrix. Indegree and outdegree are used to study several structural aspects of meaning networks. Network analyses were performed with the programme UCINET (Borgatti et al. 2002).

In Sect. 4.3.1 an adapted conceptual framework was presented in which manifest and latent functions replaced the categories of consequences and values in the original model. It was argued that this adapted model is more flexible, because manifest and latent functions are more general and inclusive categories which means that what might be a manifest functions for one individual can be a latent function for another individual. In general this implies that a particular meaning may be both a manifest and a latent function. However, we can determine the levels of both manifestness and latentness of a meaning. The level of manifestness of a meaning is defined as the ratio of outdegrees over the sum of indegrees and outdegrees of a particular meaning. The level of manifestness ranges from 0 to 1, and the higher the index the larger the proportion of times the meaning is the origin of a connection with other meanings. The level of latentness of a meaning is defined as the ratio of

Table 4.3 Adjacency matrix for the ten meanings associated with the dwelling feature number of rooms

	Multi-functionality	Family	Space	Furnishings	Tradition	Comfort	Privacy	Social contacts	Clean	Well being	Outdegrees of the row meanings
Multi-functionality	–	4	5	1		5	8	2	2	2	29
Family		–	1		1	3	2	1		1	9
Space	1		–	3	1	1	1	2	1	2	12
Furnishings	1		1	–		2			1	2	7
Tradition					–		1			1	2
Comfort	1	1	1	2		–		2	2	1	10
Privacy	2					1	–		1	2	6
Social contacts				1			1	–			2
Clean									–	1	1
Well being										–	0
Indegrees of the column meanings	5	5	8	7	2	12	13	7	7	12	78

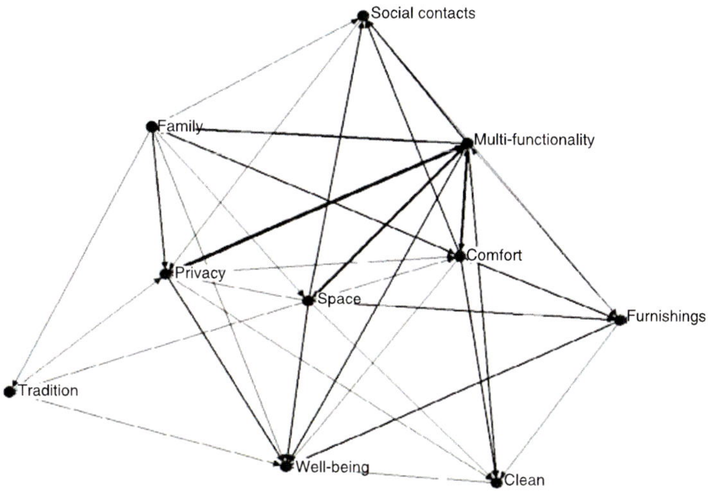

Fig. 4.5 Meaning network for dwelling feature number of rooms

Table 4.4 Some network statistics for the meanings of the number of rooms

	Level of manifestness	Level of latentness	Centrality
Privacy	0.32	0.68	0.12
Furnishings	0.50	0.50	0.09
Multi-functionality	0.85	0.15	0.22
Well-being	0.00	1.00	0.08
Comfort	0.45	0.55	0.14
Space	0.60	0.40	0.13
Clean	0.13	0.88	0.05
Contacts	0.22	0.78	0.06
Tradition	0.50	0.50	0.03
Family	0.64	0.36	0.09

indegrees over the sum of indegrees and outdegrees of a particular meaning. This index also ranges from 0 to 1, and the higher the index the larger the proportion of times the meaning is the destination of a connection with other meanings. For each meaning the sum of level of manifestness and level of latentness is 1. For the meanings of the dwelling feature number of rooms both indices are presented in Table 4.4. Multifunctionality, family and space turn out to be mainly manifest functions, while well-being, clean, contacts and privacy are to a large extent considered as latent functions.

A meaning structure of a dwelling feature is a representation of the meanings of this feature as perceived and conceived by an individual. As such it might be highly idiosyncratic, representing mainly personal meanings. It may also be less idiosyncratic in the sense that it contains meanings that are shared by other people.

Because a dwelling is considered to be, at least partly, a cultural artefact (Rapoport 1969, 1990), one might expect that the meaning structures of dwelling features contain both idiosyncratic and shared meanings. One way of investigating whether some meanings are shared more than others is by studying the centrality of meanings in a meaning network such as the one in Fig. 4.5. Although this figure seems to indicate that some meanings are more central than others, we have to be careful with our conclusions since this may be the result of the way the graphical display is constructed. A centrality measure based on indegrees and outdegrees is therefore used. The centrality of a meaning is defined as the ratio of indegrees plus outdegrees of a particular meaning over twice the sum of all entries in the adjacency matrix (cf. Wasserman and Faust 1994). Centrality ranges from 0 to 1; the higher the index, the larger the proportion of links in the meaning network that run through the particular meaning. The centrality measures for the meanings of the dwelling feature number of rooms are also depicted in Table 4.4.

Inspection of Table 4.4 shows that multi-functionality is by far the most central meaning in the meaning network for the feature number of rooms; comfort, space and privacy are also relatively central meanings in this meaning network.

It was argued in Sect. 4.2.3 that the results of a means-end analysis as represented in a hierarchical value map may be used for both marketing communication as well as for development and design purposes by, for instance, real estate agents, builders and developers. The hierarchical value map mainly allowed a qualitative analysis. By using network analysis one has not only a reliable method for presenting the relationships between the meanings, but one can also study the structural relationships between the meanings in a more quantitative way. In Table 4.4 the focus is on indices for manifestness, latentness, and centrality of meanings. Given the purpose of the analysis one might use one of these indices to focus for instance the marketing communication on the more prominent meanings.

4.4 Some New Developments

In the means-end approach as described in Sect. 4.2 the data are collected in a very open-ended way. A consequence of this is that the data collection and processing is laborious and that because of this often only a limited number of interviews is performed. The process remains laborious, even in the meaning structure approach in which the elicitation of salient dwelling features is much less open-ended and in which the meaning structures are constructed on paper during the interview. Recently, we have administered laddering interviewing in a computer-aided telephone survey, of which some results are presented in Coolen (2008) and many more in Meesters (2009). In this survey the number of features was limited and the semi-structured interviewing format was again adapted. Based on previous research only eight important dwelling features were part of the questionnaire: tenure, number of rooms, size of living room, dwelling type, garden, type of neighborhood, type of location, and type of architecture. For each of these features respondents were first asked what level they preferred. Having indicated their preferred level the respondents were subsequently asked what the most important reason was for preferring this level. This was an open-ended question, while for the coding of the answers 'field coding' was applied, where the interviewer is supplied with a set of categories within which he/she has to try to code the answer given by the interviewee. If an answer cannot be coded into one of the supplied categories, the interviewer has to note down the answer of the respondent. The set of categories for the survey was compiled on the basis of several pilot projects in which semi-structured face-to-face interviews were conducted without previously determined categories, which were subsequently transcribed and content-analyzed. The interviewers who conducted the survey were trained in field-coding the answers to the open-ended question. After the survey it turned out that between 80% and 85% of the answers were coded in one of the pre-specified categories. A content analysis was performed on the other answers, which resulted in very few additional categories. The remaining answers were too idiosyncratic to be categorized and were collected in the category 'other'. It is evident that the pilot studies were instrumental in achieving these results in the survey.

In the previous sections it has been indicated that there are several reasons for replacing the hierarchical value map by a network representation. The dwelling is a too complex and heterogeneous good to make the paper-and-pencil technique for constructing hierarchical value maps feasible. Moreover, as suggested by Valette-Florence and Rapacchi (1991), the paper-and-pencil method can lead to various omissions and errors. In addition, Van Rekom and Wierenga (2007) have recently shown that meaning structures are not necessarily hierarchical in nature, and that for non-hierarchical meaning structures a network representation is more appropriate. Up to now we have only presented meaning networks of separate housing features. Hartig (2006) has remarked that it would be interesting to integrate and aggregate the meaning networks of different housing features. Although this has not fully materialized yet, several ways of aggregating meaning networks related to housing features have been explored by Meesters (2009).

4.5 Conclusion

In this chapter the meaning structure method for measuring housing preferences has been presented. The approach focuses on *what* preferences people have and *why* they have these preferences. Because of this double focus the meaning structure method differs from most other approaches to measuring housing preferences, which only focus on *what* people want and ignore the *why* question. The meaning structure approach consists of a conceptual model, a measurement procedure, and an analysis method. The original means-end approach, as is usually applied for marketing and advertising purposes, was presented in Sect. 4.2 together with the results of a pilot study in the domain of housing. This pilot study gave rise to a revision of the conceptual framework, the measurement procedure, and the analysis method. The meaning structure method was described in Sect. 4.3. This method makes, among other things, the presentation of meaning structures more reliable by using network displays and adds a quantitative dimension to the originally more qualitatively focused analysis of these aggregated presentations by employing network statistics.

References

Bettman, J. R. (1979). *An information processing theory of consumer choice*. Reading, MA: Addison-Wesley.

Borgatti, S. P., Everett, M. G., & Freeman, L. C. (2002). *UCINET*. Columbia: Analytic Technology.

Clark, W. A. V., & Dieleman, F. M. (1996). *Households and housing. Choice and outcomes in the housing market*. New Brunswick, NJ: Center for Urban research Policy.

Cohen, J. B., & Warlop, L. (2001). A motivational perspective on means-end chains. In T. J. Reynolds & J. C. Olson (Eds.), *Understanding consumer decision making. The means-end approach to marketing and advertising strategy* (pp. 389–413). Mahwah: Erlbaum.

Coolen, H. (2006). The meaning of dwellings: An ecological perspective. *Housing, Theory and Society, 23*, 185–202.

Coolen, H. (2007). Measurement and analysis of less-structured data in housing research. *Open House International, 32*, 55–67.

Coolen, H. (2008). *The meaning of dwelling features. Conceptual and methodological issues*. Amsterdam: IOS Press.

Coolen, H., & Hoekstra, J. (2001). Values as determinants of preferences for housing attributes. *Journal of Housing and the Built Environment, 16*, 285–306.

Grunert, K. G., & Grunert, S. D. (1995). Measuring subjective meaning structures by the laddering method: Theoretical considerations and methodological problems. *International Journal of Research in Marketing, 12*, 209–225.

Gutman, J. (1982). A means-end chain model based on consumer categorization processes. *Journal of Marketing, 46*, 60–72.

Handwerker, W. P., & Borgatti, S. P. (1998). Reasoning with numbers. In H. R. Bernard (Ed.), *Handbook of methods in cultural anthropology* (pp. 549–594). Walnut Creek: Altamira.

Hartig, T. (2006). Functional bases for meanings of dwellings: Home, alone? *Housing, Theory and Society, 23*, 216–219.

Kelly, G. A. (1955). *The psychology of personal constructs*. New York: Norton.

Lundgren, B. A., & Lic, T. (2010). Customers' perspective on a residential development using the laddering method. *Journal of Housing and the Built Environment, 25*, 37–52.

Meesters, J. (2009). *The meaning of dwelling: a structural approach in people-environment relations*. Amsterdam: IOS Press.

Miles, M. B., & Huberman, A. M. (1994). *Qualitative data analysis. An expanded sourcebook* (2nd ed.). Thousand Oaks, CA: Sage.

Rapoport, A. (1969). *House form and culture*. Englewood Cliffs, NJ: Prentice-Hall.

Rapoport, A. (1988). Levels of meaning in the built environment. In F. Poyatos (Ed.), *Cross-cultural perspectives in nonverbal communication* (pp. 317–336). Toronto: C.J. Hogrefe.

Rapoport, A. (1990). *The meaning of the built environment* (2nd ed.). Tucson: University of Arizona Press.

Reynolds, T. J., & Gutman, J. (1988). Laddering theory, method, analysis, and interpretation. *Journal of Marketing Research, 28*, 11–31.

Reynolds, T. J., & Olson, J. C. (Eds.). (2001). *Understanding consumer decision making. The means-end approach to marketing and advertising strategy*. Mahwah: Erlbaum.

Reynolds, T. J., Dethloff, C., & Westberg, S. J. (2001). Advancements in laddering. In T. J. Reynolds & J. C. Olson (Eds.), *Understanding consumer decision making. The means-end approach to marketing and advertising strategy* (pp. 91–119). Mahwah: Erlbaum.

Rokeach, M. J. (1973). *The nature of human values*. New York: Free Press.

Rosch, E. (1978). Principles of categorization. In E. Rosch & B. B. Lloyd (Eds.), *Cognition and categorization* (pp. 27–48). Hillsdale: Erlbaum.

Ryan, G. W., & Bernard, H. R. (2000). Data management and analysis methods. In N. K. Denzin & Y. S. Lincoln (Eds.), *Handbook of qualitative research* (pp. 769–802). Thousand Oaks, CA: Sage.

Schwartz, S. H. (1992). Universals in the content and structure of values: theoretical advances and empirical tests in 20 countries. In M. P. Zanna (Ed.), *Advances in experimental social psychology* (Vol. 25, pp. 1–65). San Diego, CA: Academic.

Schwartz, S. H. (1994). Are there universal aspects in the structure and contents of human values? *Journal of Social Issues, 50*, 19–45.

Schwartz, S. H. (1996). Value priorities and behavior: Applying a theory of integrated value systems. In C. Seligman, J. M. Olson, & M. P. Zanna (Eds.), *Values: The Ontario symposium* (pp. 1–25). Hillsdale, MI: Erlbaum.

Valette-Florence, P., & Rapacchi, B. (1991). Improvements in means-end chain analysis: Using graph theory and correspondence analysis. *Journal of Advertising Research, 31*, 30–45.

Van Rekom, J., & Wierenga, B. (2007). On the hierarchical nature of means-end relationships in laddering data. *Journal of Business Research, 60*, 401–410.

Vyncke, P. (2002). Lifestyle segmentation: From attitudes, interests and opinions, to values, aesthetic styles, life visions and media preferences. *European Journal of Communication, 17*, 445–463.

Walker, B., Celsi, R., & Olson, J. (1987). Exploring the structural characteristics of consumer knowledge. In M. Wallendorf & P. F. Anderson (Eds.), *Advances in consumer research* (pp. 17–21). Provo: Association for Consumer Research.

Wasserman, S., & Faust, K. (1994). *Social network analysis. Methods and applications*. Cambridge: Cambridge University Press.

Williams, R. M. (1979). Change and stability in values and value systems: A sociological perspective. In M. J. Rokeach (Ed.), *Understanding human values: Individual and societal* (pp. 15–47). New York: Free Press.

Chapter 5
The Multi-attribute Utility Method

Sylvia J.T. Jansen

5.1 Introduction

When choosing between alternative places of residence, the decision-maker has to consider multiple attributes of the available alternatives at the same time, such as the preferred dwelling type, number of rooms and costs. Thus, the decision problem has multiple value dimensions, which may be in conflict, as is usually the case with difficult choices (Von Winterfeldt and Edwards 1986, p. 259). For example, the current dwelling might be relatively large and cheap but situated in a bad neighborhood, whereas an alternative dwelling might be situated in a better neighborhood but this comes at the cost of higher rent or less space. Should the resident move? Multi-criteria decision-making techniques can be used to facilitate such complex decisions. Within a multi-criteria context, decision-making problems are represented as a decision-maker who considers a set of alternatives and seeks to make an optimal decision considering all the factors (so-called criteria or attributes) that are relevant to the decision. One way of doing this is by measuring the decision-maker's values separately for a set of influential attributes and by weighting these by the relative importance of these attributes as perceived by the decision-maker. It is assumed that the more important attributes will have a greater impact in determining preferences or choices. Thus, given the factors we care about, what's the best choice? Combining the importance that respondents assign to different attributes with their evaluation of those attributes can be achieved using Multi-Attribute Utility Theory, often referred to as MAUT.

Formally, Multi-Attribute Utility theory is a technique to support decision-making when a decision-maker has to choose from a limited number of available alternatives. For example, these alternatives could be dwellings that are available at

S.J.T. Jansen (✉)
OTB Research Institute for the Built Environment,
Delft University of Technology, Delft, The Netherlands
e-mail: s.j.t.jansen@tudelft.nl

S.J.T. Jansen et al. (eds.), *The Measurement and Analysis of Housing Preference and Choice*, DOI 10.1007/978-90-481-8894-9_5,

a particular date in a particular region. The first application described in this chapter provides a simple example of such a decision-making situation. However, the method can also be used to explore respondents' preferences in a more general way in order to predict preferences and choices. For each dwelling profile, consisting of a particular combination of attribute levels, the multi-attribute utility can be calculated. Thus, dwelling profiles that are deemed to be of interest can be compared with regard to their multi-attribute utilities. Furthermore, the impact of changing a particular attribute level in terms of multi-attribute utilities can be explored. For example, keeping all other attributes constant, what is the additional value of one extra room? The second application described in this chapter provides an example of such an approach.

The goal of the present chapter is, firstly, to describe Multi-Attribute Utility theory in more detail; secondly, to examine the previous use of Multi-Attribute Utility theory in the domain of housing preferences; and, thirdly, to provide a simple and practical example of its potential application for measuring housing preferences.

5.2 Multi-Attribute Utility Methodology and Techniques

Important concepts of the theory are described in Table 5.1. With Multi-Attribute Utility theory, the overall evaluation of an alternative is defined as a weighted addition of its values with respect to its relevant attributes. This technique requires the decision-maker to evaluate the alternatives on each value dimension (called attribute) separately. For example, apartment is an attribute level of the attribute type of dwelling. Next, the decision-maker assigns relative weights to the various attributes that express the trade-off between attributes. Values and weights are then combined and aggregated by means of a formal model that generates an overall evaluation of each alternative (Von Winterfeldt and Edwards 1986, p. 6). The linear additive preference function is mostly used but other functions are also possible (see, for example, Keeney and Raiffa 1976). Important contributors to this field of research are Keeney and Raiffa (1976) and Von Winterfeldt and Edwards (1986).

Although the practical application of a multi-attribute utility method may vary, all such procedures include the following steps (Von Winterfeldt and Edwards 1986, p. 273):

1. Defining alternatives and value-relevant attributes;
2. Evaluating each alternative separately on each attribute;
3. Assigning relative weights to the attributes;
4. Aggregating the weights of attributes and the single-attribute evaluations of alternatives to obtain an overall evaluation of alternatives;
5. Perform sensitivity analyses and make recommendations.

Table 5.1 Important concepts in Multi-Attribute Utility theory

Concept	Description
Alternatives	Options where the decision-maker has to choose from, for example, various available dwellings.
Attributes	Important ('salient') characteristics of the alternatives, for example, "dwelling type" and "number of rooms".
Attribute levels	Levels of the attributes. For example, "2 rooms" is a level of the attribute "number of rooms".
Attribute value	The numerical value that is attached to a particular attribute level. A higher value is generally related to more attractiveness.
Importance score	A numerical value that indicates the importance of each attribute. A higher score is generally related to more importance.
Weight	The importance score after transformation such that, for each respondent, all attribute weights add up to one.
Single-attribute utility	The numerical strength of preference of an attribute level. It results from the multiplication of the attribute value with the attribute weight.
Combination rule	The rule that is used to aggregate over the single-attribute utilities. Usually, the simple additive rule is applied: the single-attribute utilities are simply added to obtain the multi-attribute utility.
Multi-attribute utility	The numerical strength of preference of an alternative. It results from the aggregation of single-attribute utilities.

According to Von Winterfeldt and Edwards (1986), all approaches are identical in steps 1 and 5, but may differ in the procedures for eliciting single-attribute evaluations and weights and in the models for aggregation. In the next section, these steps will be described in more detail.

5.2.1 *Step 1: Define Alternatives and Value-Relevant Attributes*

The first step in the analysis is to determine the available alternatives and their most salient attributes. For example, in the case of housing these alternatives could be available dwellings in a specific region. Note that alternatives need not necessarily be objects, but could almost be everything as long as it concerns a decision problem at hand. Next, salient attributes have to be chosen on which the alternatives will be judged. The set of attributes has to be complete, operational, decomposable, non-redundant and minimal (Keeney and Raiffa 1976, p. 50). Complete means that all important aspects of the decision problem have to be covered. The term operational is used to indicate that the attributes must have the ability to be meaningfully used in the analysis. Decomposable refers to simplifying aspects of the evaluation process by breaking it down into parts. Non-redundant means that no double counting of aspects should take place. Finally, the number of attributes should be kept as small as possible in order to prevent factors such as boredom, fatigue and confusion. The identification of the right attributes is a very important step in the procedure. Unfortunately, as Keeney and Raiffa (1976, p. 64) point out, these attributes are not simply handed to the researchers in an envelope at the beginning of the study. It requires a thorough search into the aspects that are most important to the particular decision problem. Tools to find salient attributes can be, for example, face-to-face interviews with experts, focus group interviews with experts, a literature search and methods such as Decision Plan Nets and the Repertory Grid method.

In choosing the salient attributes, Von Winterfeldt and Edwards (p. 220) advise constructing a scale that represents some natural quantitative attribute of the evaluation object, a so-called natural scale. Examples of such natural scales in the domain of housing are the size of the backyard in meters and the number of rooms. When a natural scale is not possible or available, a qualitative scale can be constructed that defines the attribute, its endpoints and possibly some intermediate marker points using verbal descriptions (for example: a quiet/lively/busy neighborhood). It may be useful to associate numbers with qualitative descriptions that are at least ordinally related to the decision maker's preferences for the levels of the qualitative scale.

However, even a natural scale may not be a satisfying value scale. The relationship between the natural scale (for example, 2, 3, 4 rooms) and the value scale (for example, 2 rooms = good, 3 rooms = better, 4 rooms = best) may not be linear or may even be non-monotone. A monotone natural scale has a value that increases or decreases monotonically with it, such that more is always better or always worse (note that the relationship need not be perfectly linear). Monotonicity can be violated for some attributes. For example, there may be a maximum to the preferred number of rooms. Above some maximum number the added value may become negative as more space may not compensate for more maintenance.

Value functions may be linear in the sense that the distance in value between consecutive attribute levels is about equal. However, value functions can also be concave or convex. For example, the value function of the attribute number of rooms may be concave as the added value of each additional room may be worth less than the previous one (a decreasing marginal evaluation, see, for example, Keeney and Raiffa 1976, p. 88). This means that the step from four to five rooms may be appreciated more than the step from five to six rooms, and so forth. Von Winterfeldt and Edwards (1986, p. 237) advise, in cases where concave or convex functions are observed, reformulating the evaluation problem in order to produce linear or near-linear value functions by carefully selecting or creating a scale. This can be done, for example, by setting limits to the minimum and maximum attribute levels to be evaluated. The argument for this is that a value function is linear in the natural scale that most closely reflects the value concerns to which it is related. For example, to prevent decreasing added value, a maximum can be set on the number of rooms in order to obtain an interval with a more or less linear relationship between value and size.

Thus, one must carefully select and construct the attribute scales that are used to explore the attribute values. However, even after careful selection and the construction of natural scales, value functions may not be perfectly monotone linear. In such cases, curve fitting procedures can be used, such as exponential or polynomial functions, to estimate the attribute level values. After fitting such a function, values at scale points for which no assessments were made, can be interpolated.

5.2.2 *Step 2: Evaluate Each Alternative Separately on Each Attribute*

After determining the salient attributes, the alternatives have to be evaluated for each of these attributes. For example, what is the value of an alternative consisting of a dwelling with three rooms? And with four rooms? Von Winterfeldt and Edwards (1986, chapters 7 and 8) present a variety of methods that can be used to elicit single-attribute value functions and utility functions. Here, some often-used methods will be described.

With direct rating, the respondent considers at least three stimuli: two stimuli that are used as endpoints or anchors and one that is used to elicit the numerical judgment. A "bad" or least preferred stimulus is arbitrarily assigned the value of 0 and a "good" or most preferred stimulus has been given the value of 100. The intermediate stimulus (or stimuli) is judged in relation to these anchor points. The relative spacing between points reflects the strength of one stimulus over another. For example, if a dwelling with one room is used as the lower anchor (0) and a dwelling with six rooms is used as the higher anchor (100), the value of a dwelling with two, three, four or five rooms can be determined between these extremes.

With the difference standard sequences technique the respondent identifies a sequence of stimuli that are equally spaced in value. For example, backyard length of 5, 10, 13 and 15 m may be equivalent to a value of 0.25, 0.50, 0.75 en 1.0, respectively. With the bisection method, the most and least preferred option are identified and a midpoint stimulus is found that is equidistant in value from both extremes. For example, the respondent is asked which number of rooms is least preferred (value = 0) and most preferred (value = 100). Next, the number of rooms that lies in value exactly between these anchors is determined.

5.2.3 *Step 3: Assign Relative Weights to the Attributes*

After evaluating all alternatives on all salient attributes, the importance of each of the attributes is determined and the weights are calculated. A number of techniques is available to determine the importance of the attributes. See, for example, Von Winterfeldt and Edwards for an overview (1986, chapter 8, p. 274). Well-known methods are ranking, direct rating, ratio estimation and the method of swing weights. With ranking, the respondent is asked to rank all attributes in order of importance. An example of direct rating is to divide 100 points over the attributes so that the number of points reflects the relative importance of the attributes. For the ratio estimation method, first the least important attribute is determined.

Next, the respondent is asked how much more important each attribute is when related to the least important one. Finally, for swing weighting, the respondent is asked how much an attribute contributes to the overall value of the alternatives relative to other attributes. Usually with this method the respondent is provided with a choice between profiles reflecting the worst and best levels of each attribute. The respondent is asked to indicate which of the differences between the worst and the best level (called swings) contributes most in overall value. Then, the extent to which the value swings differ between attributes is assessed by letting the respondent assign a score to the relative significance of the range when compared to the most important range. For example, a profile consisting of a dwelling with two rooms and a backyard of 5 m (worst levels) is compared to a profile consisting of a dwelling with five rooms and a backyard of 15 m (best levels). The respondent first indicates that the swing from two to five rooms is more important than the swing from a backyard of 5–15 m in length. Next, the respondent indicates that this swing is four times less important than the swing from two to five rooms.

After scores have been collected for the importance of the attributes, these are transformed into weights by dividing, for every respondent, the rating of each attribute by the sum of all ratings (Von Winterfeldt and Edwards 1986, p. 281):

$$w_i = \frac{w_i^{'}}{\sum_{i=1}^{n} w_i^{'}}, \tag{5.1}$$

where $w_i^{'}$ is the not-normalized ratio weight and w_i the normalized weight. Hereby individual weights for each attribute are obtained that add to 1, as is conventional in Multi-Attribute Utility theory (Von Winterfeldt and Edwards 1986). Assume, for example, that a respondent has the following importance scores for eight attributes: 20, 30, 40, 50, 30, 60, 70, 20. The sum of these ratings is 320. The weight for the first attribute is therefore: 20/320 = 0.06. The other weights are calculated in the same way.

5.2.4 *Step 4: Aggregate the Weights of Attributes and the Single-Attribute Evaluations of Alternatives to Obtain an Overall Evaluation of Alternatives*

Weights and single-attribute values or functions can be aggregated using a variety of models. The weighted linear additive preference function is the most commonly applied aggregation method (Von Winterfeldt and Edwards 1986, p. 275). Other models are, for example, the multiplicative model and the multi-linear model. These models are not frequently used. With the weighted linear additive function, the overall evaluation of an alternative is calculated by multiplying the weight by the attribute value for each attribute and summing these weighted attribute values over all attributes

(Payne et al. 1993). It is assumed that the alternative with the highest overall evaluation will be chosen. The weighted linear additive preference function processes all of the relevant information. Furthermore, the conflict among values is assumed to be addressed and resolved by explicitly considering the extent to which one is willing to trade off attribute values, as reflected by the relative importance or weights. This simple preference function is a compensatory combination rule as a low value of one attribute can, at least partially, be compensated by higher values on one or more of the remaining attributes. The multi-attribute utility for alternative x is:

$$v(x) = \sum_{i=1}^{n} w_i v_i(x_i), \tag{5.2}$$

where $v_i(x_i)$ is the value of alternative x on the ith attribute, w_i is the importance weight of the ith attribute, and n is the number of different attributes (Von Winterfeldt and Edwards 1986, p. 263, p. 275).

In theory, there are various techniques to elicit values, they can be combined with different techniques for calculating weights and they can be aggregated with a number of models. Thus, the application of Multi-Attribute Utility theory can differ considerably between studies. However, in practice, a very limited combination of techniques is used. The most important procedure is termed SMART (Simple Multi-Attribute Rating Technique) (Edwards and Newman 1982). This procedure is simple and easily applicable; it consists of the direct rating technique for eliciting values combined with the ratio estimation technique for calculating weights and the weighted linear additive preference function.

5.2.5 *Step 5: Perform Sensitivity Analyses and Make Recommendations*

In the fourth step, multi-attribute utilities for all alternatives have been calculated. The alternative with the highest multi-attribute utility should be the preferred choice. In the last step of the procedure, sensitivity analyses are carried out to evaluate the stability of the results. The impact of different values and weights on the multi-attribute utilities of the available alternatives can be determined. One way to obtain different values and weights is by using different elicitation methods. For example, both the direct rating technique and bisection method could be used to obtain values. Multi-attribute utilities could then be calculated twice: once using the values obtained with the direct rating technique and once with the values obtained with the bisection method. The resulting multi-attribute utilities for the available alternatives can be compared and the robustness of the results can be determined.

Another way to obtain insight into the stability of the results is to use a different weighting technique. For example, the equal weight function can be applied. This technique simplifies the choice process by ignoring information about the relative importance of each attribute (Jia et al. 1998; Bettman et al. 2006, p. 329). In doing so the method assumes that all attributes have equal weight. An overall value for each alternative is obtained by simply summing the values for each attribute for that alternative. For example, for eight attributes, the weight of each attribute is 1/8 = 0.125. Other weighting techniques include the rank reciprocal rule and the rank sum weighing procedure (Von Winterfeldt and Edwards 1986).

In this section Multi-Attribute Utility theory has been explained in more detail. Note that this was just a short description as Multi-Attribute Utility theory is a method that has been developed and used in many research domains and can be quite complicated. Interested readers are referred to, for example, Keeney and Raiffa (1976) and Von Winterfeldt and Edwards (1986). Also, computer programs have been developed to help decision-makers in structuring the problem. In the next section, some studies that have used the Multi-Attribute Utility method in the field of housing will be discussed.

5.3 Previous Research in the Domain of Housing Using Multi-Attribute Utility Theory

Although Multi-Attribute Utility theory has been widely used in all kinds of research domains, it has not been applied frequently in the domain of housing. Veldhuisen and Timmermans (1984) used a multi-attribute method – that they termed magnitude estimation – to measure preferences for individual housing attributes and to identify overall evaluations. They compared the method with two other methods, conjoint measurement and functional measurement. They concluded that all three measurement methods might be used to obtain both single-attribute and overall utilities. However, they showed that a multiplicative preference function might perform slightly better than an additive preference function. Lindberg et al. (1989) used an approach based on Multi-Attribute Utility theory. These authors assumed that a person's evaluation of a given housing alternative is determined by a combination of the evaluation of its different attributes (conform Multi-Attribute Utility theory). In addition, they assumed that a person's evaluation of each housing attribute is determined by what effect he/she believes it to have on the possibilities to attain various life values and his/her evaluations of those life values. Ultimately, the authors concluded that a model assuming that the evaluation of a given alternative is determined by a weighted sum of the evaluations of the attained life values, was found to be the most successful one in predicting both preference ratings and choices, without specifying how individual attributes contribute to this value-fulfillment.

Breij et al. (1989) describe a computer program called MIDAS (Multi-attribute Individual Decision Assistance System). This program is developed as a general program and is, among other domains, applied in the domain of housing. The program allows the respondent to choose the salient attributes and attribute levels. The evaluation of the attribute levels and the importance of the attributes is determined as well as the ideal attribute level. The computer then calculates the best option on the basis of the responses provided by the respondent. Breij and coauthors report on a study performed with MIDAS into the housing preferences of young professionals. They concluded that in 84% of cases respondents agreed with the best alternative provided by MIDAS that was based on their own evaluation and importance ratings. Half of the respondents, who did not agree with the proposed alternative, argued that the proposed option was too expensive. The authors were satisfied with the functioning of the computer program to elicit housing preferences. However, whether the program is currently used in the domain of housing is not known. An Internet search using Google did not lead to any results.

Vreeker (2006) used two multi-criteria evaluation methods, including Multi-Attribute Utility theory, to evaluate four alternatives for the Amsterdam South Axis urban development program in the Netherlands. He compared the two multi-criteria evaluation methods as well as the results they produced. The focus was on the information they provide to the decision-maker. The effect of the alternatives on seven indicator themes, such as traffic and urban quality, was assessed. The results of both methods matched with regard to the most and least preferred alternative but not for the second and third preferred alternative. Vreeker concluded that the robustness of the outcomes should always be checked for dependence upon the applied method. However, he also concluded that both methods agreed on which alternative would no longer be considered to be a serious alternative (the option with the least multi-attribute utility).

In a domain related to housing, Canbolat et al. (2007) combined the decision tree technique and Multi-Attribute Utility theory for selecting a country for a new plant (an automotive brake manufacturing facility). The authors concluded that the software program that was developed based on Multi-Attribute Utility theory enabled the decision-maker to clearly understand the magnitude of the relative strengths and weaknesses of the various alternatives.

Domains in which Multi-Attribute Utility theory is frequently used are, for example, water control and environmental studies. Examples of recent studies are: ranking irrigation subsystems in India (Raju and Vasan 2007), reconsidering heterogeneity and aggregation issues in environmental valuation (Barreiro-Hurlé and Gómez-Limón 2008), estimating the potential impacts of irrigation water pricing (Latinopoulos 2008) and flood risk mapping exemplified at the Mulde river in Germany (Meyer et al. 2009). Linkov et al. (2006) provide an overview of 20 recent (<10 years) Multi-Attribute Utility studies in the field of environmental management.

5.4 Practical Application of Multi-Attribute Utility Theory in the Domain of Housing

The following part of this chapter is dedicated to describing two examples of the use of Multi-Attribute Utility theory in the field of housing. The first is a hypothetical example of a decision-making problem for choosing a new place of residence. The second example is based on a real research case in which the preference-structure of respondents with at least an average income and living in the Netherlands is examined.

5.4.1 Example 1: Decision Analysis Example

A household consisting of two parents (Paul and Laura) and three small daughters decides to search for a new dwelling because their current dwelling doesn't have enough room since the birth of their third daughter. The couple requires an owner-occupied home with a backyard for their small children. After a thorough search with the use of some internet sites proposing available dwellings, the couple finds out that there are six available dwellings within their selected region that satisfy their requirements. These dwellings differ with respect to their characteristics (so-called attributes).

The couple decides to examine which alternative they should choose using a Multi-Attribute Utility method. Firstly, they decide which attributes are important to them. After some deliberation, they come up with the following ones: dwelling type, costs, size of the living room, number of rooms, size of the backyard, architectural style and residential environment. The six available dwellings can be described in terms of their attribute levels. These descriptions are termed dwelling profiles. They are presented in Table 5.2.

Next, they determine the values of the attributes of each alternative using a rating scale with two anchors. On the left side the rating scale is anchored by "extremely unattractive" (0) and on the right side by "extremely attractive" (100). Paul and Laura's individual values, as well as their mean values, are presented in Table 5.3. For example, Laura has provided a value of 50 for a ground floor apartment and feels more attracted to a semi-detached house, which she has given a value of 90. In contrary, Paul likes the ground floor apartment most (value of 90) and finds the semi-detached house the least attractive (a value of 40). The table shows that the couple each assigns quite different values. They do agree about the least preferred and most preferred characteristics with regard to size of the living room, number of rooms and residential environment. But their evaluation differs with regard to the other attributes. This justifies a decision-analysis method.

Next, the couple decides to assign a score to each of the attributes according to their importance. They use a rating scale with numerically scaled endpoints of 0 (not important at all) and 100 (extremely important). Thus, the higher the importance, the higher the score for that particular attribute. The importance scores are

Table 5.2 Six dwelling profiles on the basis of their attribute levels

	Dwelling type	Costs	Size living room	Number of rooms	Size backyard	Architectural style	Residential environment
A	Semi-detached	€ 220,000	20 m^2	2 rooms	5 m	Innovative	Rural
B	Semi-detached	€ 140,000	20 m^2	4 rooms	15 m	Traditional	Sub-urban
C	Terraced/corner	€ 300,000	40 m^2	4 rooms	10 m	Traditional	Rural
D	Terraced/corner	€ 140,000	40 m^2	2 rooms	15 m	Innovative	Urban
E	Apartment	€ 220,000	20 m^2	3 rooms	10 m	Traditional	Rural
F	Semi-detached	€ 220,000	30 m^2	2 rooms	10 m	Modern	Urban

Table 5.3 Attribute values

	Laura	Paul	Couple
Dwelling type			
Apartment (ground floor)	50	90	70
Terraced house/corner house	70	60	65
Semi-detached house	90	40	65
Purchase costs			
€ 140,000	40	50	45
€ 220,000	70	100	85
€ 300,000	60	25	42
Size living room			
20 m²	70	40	55
30 m²	90	70	80
40 m²	100	100	100
Number of rooms			
2	60	20	40
3	70	80	75
4	80	100	90
Backyard size			
5 m	70	100	85
10 m	90	50	70
15 m	100	10	55
Architectural style			
Traditional	80	90	85
Innovative	50	20	35
Modern	80	70	75
Residential environment			
Urban	60	10	35
Sub-urban	80	60	70
Rural	90	100	95

presented in Table 5.4. After that, they normalize the importance ratings into weights by dividing the rating of each attribute by the sum of all the ratings. The weights obtained in this way sum up to one. For example, for Laura the importance score for dwelling type is 95 and the sum of all importance scores is 625. The weight for dwelling type thus becomes 95/625 = 0.15. The weights are provided in Table 5.4.

For Laura, all attributes are almost equally important; they range from 0.13 to 0.16. The weights assigned by Paul show more variation, he finds the size of the backyard relatively unimportant (weight = 0.08), but the costs and number of rooms are very important to him (weight = 0.22).

Finally, the couple multiplies the values with the weights for every attribute level to calculate single-attribute utilities. The results are presented in Table 5.5. A higher single-attribute utility indicates that the particular attribute level is valued highly and deemed to be important. A lower utility indicates that the particular attribute level is relatively unimportant, lowly valued, or both. For Laura a backyard with a

Table 5.4 Importance scores and weights

	Importance scores			Weights		
	Laura	Paul	Couple	Laura	Paul	Couple
Dwelling type	95	55	75	0.15	0.12	0.14
Purchase costs	100	100	100	0.16	0.22	0.19
Size living room	80	55	67.5	0.13	0.12	0.12
Number of rooms	95	100	97.5	0.15	0.22	0.19
Backyard size	95	35	65	0.15	0.08	0.11
Architectural style	80	55	67.5	0.13	0.12	0.12
Residential environment	80	55	67.5	0.13	0.12	0.12
Total	625	455	540	1	1	1

Table 5.5 Single-attribute utilities

	Laura	Paul	Couple
Dwelling type			
Apartment (ground floor)	7.6	10.9	9.6
Terraced house/corner house	10.6	7.3	8.9
Semi-detached house	13.7	4.8	8.9
Purchase costs			
€ 140,000	6.4	11.0	8.5
€ 220,000	11.2	22.0	16.1
€ 300,000	9.6	5.5	8.0
Size living room			
20 m^2	9.0	4.8	6.8
30 m^2	11.5	8.5	10.0
40 m^2	12.8	12.1	12.4
Number of rooms			
2	9.1	4.4	7.4
3	10.6	17.6	13.9
4	12.2	22.0	16.7
Backyard size			
5 m	10.6	7.7	9.7
10 m	13.7	3.8	8.0
15 m	15.2	0,8	6.3
Architectural style			
Traditional	10.2	10.9	10.6
Innovative	6.4	2.4	4.4
Modern	10.2	8.5	9.3
Residential environment			
Urban	7.7	1.2	4.4
Sub-urban	10.2	7.3	8.7
Rural	11.5	12.1	11.8

size of 15 m, a semi-detached house and a backyard with a size of 10 m obtain the most utility. For Paul, purchase costs of € 220,000 and a dwelling with four rooms yield the most utility. For the couple, most utility is obtained from the attribute level of four rooms, directly followed by the attribute level of three rooms. An innovative

Table 5.6 Multi-attribute utilities aggregated with the use of two different weighting methods

	Weighted additive method			Equal weights method		
Dwelling Profile	Laura	Paul	Couple	Laura	Paul	Couple
A	71.5	58.2	65.2	71.5	60.1	65.8
B	76.9	61.5	66.6	77.2	55.8	66.5
C	80.6	73.6	76.4	81.5	75.1	78.2
D	68.2	39.1	52.3	68.6	38.6	53.6
E	73.8	82.1	76.9	74.4	78.6	76.5
F	77.1	53.2	64.1	77.2	51.5	64.3

architectural design and an urban residential environment provide the least single-attribute utility.

Finally, the couple uses the weighted linear additive preference function to calculate multi-attribute utilities. This means that the single-attribute utilities are simply aggregated according to the combination of attribute levels in the dwelling profiles. Table 5.6 shows the multi-attribute utilities for the six available dwellings in the column labeled "Weighted additive method". For the couple, the highest utility score is obtained for dwelling E, indicating that they should choose this dwelling.

The couple decides to carry out a sensitivity analysis to evaluate the robustness of the results. Firstly, they examine their individual multi-attribute utilities. Table 5.6 shows that dwelling E is the best choice for Paul, but not for Laura. For Laura, this dwelling is only fourth in preference ranking based on multi-attribute utilities. A further exploration shows that dwelling C might also be a good choice; it is the option with the highest multi-attribute utility for Laura and the second highest utility for Paul. As a second sensitivity analysis, they calculate multi-attribute utilities using the equal weights method. This means that every attribute has the same importance (in this case 0.143) and that multi-attribute utilities are calculated using these weights. The results are presented in Table 5.6 in the column labeled "Equal weights method". Dwelling C now has the highest multi-attribute utility for the couple. It is the best option for Laura and the second best option for Paul. This sensitivity analysis teaches the couple that besides dwelling E, dwelling C might also be a good alternative. It also shows the potential impact that the importance of attributes can have on the calculated best choice alternative.

5.4.2 Example Two: Calculating Single-Attribute Utilities for Dwelling Attributes in a Large Sample

The previous section described a hypothetical decision-making situation regarding the choice for a new place of residence. In this section, single-attribute and multi-attribute utilities for dwelling profiles are described that were obtained from about 2,000 respondents in a recent study. Firstly, the background of the study is described.

5.4.2.1 Background

The data were collected in the context of the large study Huizenkopers in Profiel (HIP; Boumeester et al. 2008b) that is performed every 2 years. In this study, data on housing preferences and the current housing situation were obtained from respondents who have at least a standard income. The goal of the Huizenkopers in Profiel study is to determine the needs and wishes of home buyers, including future home buyers, in order to establish what has to be built. The data were collected using telephone interviews between February and April 2008. The multi-attribute utility questions were answered by 2,047 respondents. The characteristics of the respondents are presented in Table 5.7.

5.4.2.2 Step 1. Define Alternatives and Value-Relevant Attributes

We did not select alternatives (dwelling profiles) but provided respondents with a number of pre-selected attributes and attribute levels. The set of attributes considered are those that were used in the pilot study for the VROM module 'Consumer behavior' (Consumentengedrag) (Boumeester et al. 2008a; Jansen et al. 2009). In this study, 13 attributes were chosen on the basis of expert opinions and the literature that had shown these attributes to be important attributes influencing residential decision-making (Floor and van Kempen 1994, Goetgeluk 1997; Heins 2002; Boumeester et al. 2005). As explained before, the multi-attribute utility questions were part of a

Table 5.7 Respondents' characteristics

Gender (n = 2,043)	
Female	1,087 (53%)
Age (n = 2,037)	
Mean (std)	52 (13)
Range	19–89
Number of persons in household (n = 2,038)	
One	298 (15%)
Two	896 (44%)
Three	291 (14%)
Four	368 (18%)
Five or more	185 (9%)
Children < 18 (n = 902)	
None	291 (32%)
One	221 (25%)
Two	268 (30%)
Three or more	122 (13%)
Paid job (n = 2,036)	
Yes	1,318 (65%)
Education (n = 1,979)	
Lower	474 (24%)
Middle	716 (36%)
Higher	789 (40%)

Table 5.8 Attributes and attribute levels

Dwelling type	Purchase costs/rental costs
Apartment	€ 140,000/€ 338 per month
Terraced house/corner house	€ 220,000/€ 532 per month
Semi-detached house	€ 300,000/€ 725 per month
Tenure	Size of the living room
Rental house	20 m^2
Owner-occupied house	30 m^2
	40 m^2
Architectural style	Number of rooms
Traditional	2
Innovative	3
Modern	4
Residential environment	Backyard size/size balcony
Urban	5 m/4 m^2
Sub-urban	10 m/7 m^2
Rural	15 m/10 m^2

larger survey into residents' housing preferences (Huizenkopers in Profiel: Boumeester et al. 2008b). For this reason, we could not include all 13 attributes, because the interview burden for the respondents would be too high (fatigue, boredom, no more time, and so on). Except for residential environment, only attributes were included that pertained to characteristics of the dwelling. This choice was based on previous findings that dwelling characteristics are deemed to be more important than characteristics of the dwelling environment (Boumeester et al. 2008a). Furthermore, some dwelling environment characteristics may not be easy to influence, such as the composition of the residents living in the neighborhood. However, residential environment (urban, suburban, and rural) was included in the survey, because this attribute has proven to be important to respondents (Boumeester et al. 2008a). For all attributes two to three attribute levels were chosen. The attributes and attribute levels are presented in Table 5.8. All analyses were performed using SPSS, version 14.0.

5.4.2.3 Step 2. Evaluate Each Alternative Separately on Each Attribute, That Is, Assess the Level of Satisfaction with the Attributes

The attribute level values were obtained directly with the use of rating scales. Respondents were asked to indicate their likes or dislikes with regard to each level of every attribute on a scale with two anchors: on one side "extremely unattractive" with an assigned value of 0 and on the other side "extremely attractive" with a value of 100. The questions were introduced by explaining these endpoints and stating that a higher appointed number was related to more attractiveness. Furthermore, the interviewer explained that the respondent had to take their current situation and household income as a starting point when answering the questions.

For example, when inquiring about living in a dwelling with three rooms, the following question was formulated: Please indicate on a scale anchored by 0 "extremely unattractive" and 100 "extremely attractive" how you would value living in a dwelling with three rooms. Next, the respondent was asked about his/her evaluation of living in a dwelling with four rooms, and then with five rooms. This procedure was repeated for each level of every attribute.

Table 5.9 shows the mean values for the attribute levels obtained in this study. Note that in the current study, respondents were presented with questions on either

Table 5.9 Mean values for attribute levels

	Mean	Standard deviation	Number of respondents
Dwelling type			
Apartment	40.9	34.1	2,034
Terraced house/corner house	57.4	29.4	2,034
Semi-detached house	67.1	27.5	2,033
Tenure			
Rental house	41.4	33.5	2,028
Owner-occupied house	79.3	28.1	2,031
Purchase costs			
€ 140,000	44.5	33.5	1,528
€ 220,000	59.3	28.1	1,531
€ 300,000	56.2	33.5	1,532
Rental costs			
€ 338 per month	70.4	30.0	474
€ 532 per month	53.9	29.9	473
€ 725 per month	26.7	29.1	473
Size living room			
20 m^2	25.5	24.3	2,033
30 m^2	50.1	26.6	2,032
40 m^2	70.1	25.1	2,032
Number of rooms			
2	16.6	19.6	2,032
3	41.2	28.5	2,032
4	65.3	26.7	2,032
Backyard size			
5 m	30.1	26.1	1,721
10 m	53.7	29.0	1,721
15 m	63.4	31.1	1,721
Size balcony			
4 m^2	31.8	28.7	243
7 m^2	46.3	29.1	243
10 m^2	61.9	31.4	243
Architectural style			
Traditional	76.8	19.8	2,027
Innovative	52.6	27.3	2,018
Modern	58.3	26.1	2,023
Residential environment			
Urban	43.2	29.3	2,028
Sub-urban	55.7	24.2	2,028
Rural	72.2	26.0	2,030

rental costs or purchase costs, depending upon their current living circumstances with regard to tenure. Similarly, only respondents with a backyard or a balcony were presented with questions on backyard and balcony size, respectively.

However, ultimately all respondents answered questions for the same eight attributes. The results show that respondents value the attribute level of owner-occupied houses the highest, followed by a traditional architectural style and a rural residential environment. The attribute level of two rooms was valued the least, followed by a living room size of 20 m^2 and monthly rental costs of € 725.

In general, the ranking of levels within a particular attribute is as expected, with more space and less costs related to higher preferences. An exception to this rule is purchase cost, because a dwelling of € 140,000 is not as highly valued as the more expensive houses. One would expect lower purchase costs to be evaluated more positively than higher purchase costs. This seems to suggest that respondents take other things into consideration, besides price, when evaluating the single attribute of purchase costs. For example, they might be worried about the size or the state of maintenance of such a relatively cheap dwelling. Thus, quality is implicitly included in the price. Price may reflect all kinds of qualities, such as maintenance, neighborhood amenities, image of the neighborhood, and so on. This finding agrees with that of Floor and van Kempen (1994) who notice that respondents more frequently pronounce a desire for a more expensive single-family owner-occupied dwelling than for a less expensive one. He argues that, apparently, respondents are aware that it is difficult to find a reasonable owner-occupied dwelling for a low price. Park et al. (1981) used a different method (the Decision Plan Net) but also observed an inconsistency with regard to the attribute of price. Their respondents were more than averagely satisfied with the attribute of price, even though they had paid more for their new home than they had previously found acceptable. Park and co-authors concluded that their subjects may not have perceived price as a separate dimension. When the other dimensions had appeared very satisfactory, the respondents' perceptions of price might have been adjusted to reflect the net worth of the total entity. In this study the effect is not observed for rental houses; here the highest evaluation is for the dwelling with the lowest rental price. Presumably, the respondents expect price in a regulated rental market price not to be so strongly related to the quality of the dwelling and the neighborhood.

5.4.2.4 Step 3. Assign Relative Weights to the Attributes

Next, the respondents were asked to assign an importance rating to each of the attributes on a rating scale, with numerically scaled units from 0 (not important at all) to 100 (extremely important). Ties (i.e., the same rating scores) are allowed if attributes are considered to be of equal importance. An introduction was provided explaining the endpoints of the rating scales. Subsequently, the respondents were presented with each of the eight attributes and asked to provide a rating indicating the importance of the particular attribute.

Table 5.10 Mean importance scores and weights

	Importance scores			Weights		
	N	Mean	Std	N	Mean	Std
Dwelling type	2,030	77.1	18.8	2,030	0.124	0.028
Tenure ship	2,030	78.2	22.1	2,030	0.126	0.033
Purchase costs	1,548	84.4	17.3	1,548	0.135	0.027
Rental costs	476	81.8	19.8	476	0.138	0.032
Size living room	2,031	79.6	14.4	2,031	0.129	0.021
Number of rooms	2,031	82.1	13.8	2,031	0.133	0.021
Backyard size	1,723	70.1	22.4	1,723	0.111	0.032
Size of balcony	242	66.1	26.7	242	0.105	0.038
Architectural style	2,030	70.0	21.1	2,030	0.112	0.029
Residential environment	2,032	82.4	15.1	2,032	0.134	0.031

The mean importance ratings are presented in the column labeled "Importance scores" in Table 5.10. All attributes are deemed relatively important, as the lowest mean score was 66 on a 100-point scale. This means that no superfluous attributes were chosen in this study, which makes sense given the careful selection of the attributes. The mean scores are relatively close to each other (range 66–84) indicating that the attributes are almost all equally important to the respondents. Purchase costs, residential environment and number of rooms are perceived as being the most important dwelling characteristics. Size of the balcony, backyard size and architectural style are deemed to be the least important dwelling characteristics.

The importance scores were transformed to weights as previously explained. Table 5.10 shows the mean weights in the column labeled "Weights". Interestingly, rental cost has the highest weight (thus the most importance). Apparently, respondents who have a rental home, attach relatively high importance to the rental costs in comparison to the other attributes. No other differences in the ranking of the importance scores and weights were observed.

5.4.2.5 Step 4. Aggregate the Weights and the Attribute Evaluations of Alternatives to Obtain an Overall Evaluation of Alternatives, Being the Multi-Attribute Utility

Table 5.11 shows the mean single-attribute utilities for each of the attribute levels. The utilities represent the value that is added to the subjective value of a dwelling, keeping all other attribute levels constant. Thus, a traditional architectural design would add in general 8.64 utility to the multi-attribute utility, whereas a modern and an innovative design would add only 6.41 and 5.80, respectively. Irrespective of other attribute levels, an owner-occupied house generally yields the most utility,

Table 5.11 Single-attribute utilities

	Number of respondents	Mean	Standard deviation
Dwelling type			
Apartment	2,034	5.1	4.6
Terraced house/corner house	2,031	7.0	3.9
Semi-detached house	2,030	8.3	3.8
Tenure			
Rental house	2,027	4.9	4.3
Owner-occupied house	2,030	10.0	4.6
Purchase costs			
€ 140,000	1,526	6.1	4.7
€ 220,000	1,530	8.1	4.3
€ 300,000	1,531	7.6	4.8
Rental costs			
€ 338 per month	473	9.9	4.7
€ 532 per month	472	7.4	4.5
€ 725 per month	472	3.6	4.2
Size living room			
20 m^2	2,033	3.2	3.1
30 m^2	2,031	6.4	3.6
40 m^2	2,031	9.0	3.6
Number of rooms			
2	2,033	2.2	2.7
3	2,032	5.5	4.1
4	2,032	8.7	3.9
Size of backyard			
5 m	1,723	3.3	2.9
10 m	1,721	6.1	3.6
15 m	1,721	7.4	4.0
Size of balcony			
4 m^2	242	3.3	3.2
7 m^2	242	5.0	3.5
10 m^2	242	6.7	4.1
Architectural style			
Traditional	2,027	8.6	3.2
Innovative	2,020	5.8	3.4
Modern	2,024	6.4	3.3
Residential environment			
Urban	2,028	5.7	4.0
Sub-urban	2,028	7.4	3.5
Rural	2,029	9.6	3.9

indicating that it is valued positively and deemed important. It is followed in utility by the attribute levels "rental costs of € 338 a month" and "rural residential environment".

Since the weights add to 1 for every respondent and the values are between 0 and 100, all multi-attribute utilities fall between 0 and 100. A higher multi-attribute utility score is related to more preference. Taking into account the attribute levels yielding the highest utilities, the ideal dwelling would follow this description: a semi-detached owner-occupied house with a purchase cost of € 220 000, with a living room area of 40 m^2, 4 rooms, a backyard 15 m long, in a traditional architectural style and located in a rural residential environment. The total multi-attribute utility for this dwelling is 69.8. Similarly, a rental apartment with € 725 rental costs per month, a living room surface area of 20 m^2, 2 rooms, a balcony with a surface area of 4 m^2, an innovative architectural style and which is located in an urban residential environment receives, in general, the lowest multi-attribute utility score (33.9). All other combinations of attribute levels lie between these extremes.

As each respondent evaluated eight attributes of which seven had three levels and one had two levels, the total possible number of alternatives is $3^7 * 2^1 = 4{,}374$. This number is even larger considering the fact that some respondents evaluated purchase costs and size of the backyard whereas others evaluated rental costs and size of the balcony. This method can be used to compare a selected number of dwelling alternatives as was shown in the first example presented in this chapter. The method can also be used to explore the impact of changing the level of an attribute on the attractiveness of the overall dwelling profile. For example, if there is a choice between building dwellings with larger living rooms or dwellings with larger backyards, multi-attribute utilities for these dwelling profiles can be compared keeping all other attributes constant. In the present example, a backyard with a size of 10 m instead of 15 m would incur a loss in utility of 7.4– 6.1 = 1.3. In contrast, a living room with a size of 40 m^2 instead of 30 m^2 would increase the multi-attribute utility by 9.0 – 6.4 = 2.6. Thus, in this particular example, respondents derive more utility from a larger living room than from a larger backyard and builders could use this information to optimize building decisions to clients' needs. In a similar way it can be seen that the largest increase in utility would result from building dwellings with four rooms instead of only two.

Another way in which the results obtained in this and similar studies could be used is segmentation research. Based on the particular characteristics of dwellings, for example apartments with two rooms in an urban residential environment, more or less homogeneous groups of consumers could be sought with the highest multi-attribute utilities for that particular dwelling profile. The segmentation could take place on the basis of demographic and geographic characteristics and the resulting groups could be the target groups for specific marketing activities.

5.4.2.6 Step 5. Perform Sensitivity Analyses and Make Recommendations

In the last step of the procedure, sensitivity analyses can be carried out to examine the robustness of the resulting single-attribute and multi-attribute utilities. As explained in the theoretical part of this chapter, various methods can be chosen for this.

5.5 Discussion of the Use of Multi-Attribute Utility Theory in the Domain of Housing Preferences

The goal of this chapter was to explain and to explore the use of Multi-Attribute Utility in the domain of housing preferences research. In the previous sections, two examples of Multi-Attribute Utility theory in the domain of housing were described. The first example showed a hypothetical decision-making situation. This is the case when a number of options are available and the decision-maker has to choose between them. The best choice would be the alternative with the highest multi-attribute utility. The second example showed the attractiveness and the importance of eight dwelling characteristics according to about 2,000 respondents. The method allows the calculation of the multi-attribute utility for each possible combination of dwelling characteristics. Also, the impact of varying attribute levels can be examined by replacing one (or more) attribute levels with another and calculating the difference in utility.

Note that the current study has a number of limitations. Firstly, the current application of Multi-Attribute Utility theory was somewhat loosely applied. Formally, when using the direct rating technique, the worst and best level of each attribute should be used as endpoints or anchors (Von Winterfeldt and Edwards 1986, p. 218, p. 227). These anchors should be assigned a value of 0 and 100, respectively, and the remaining stimuli should be compared to these endpoints. For example, a dwelling with one room could have been referred to as a "bad" endpoint (value 0) and a dwelling with 6 rooms as a "good" endpoint (100). The attribute levels 2, 3, 4 and 5 rooms could then have been judged relative to these anchors. However, it was considered to be too difficult for respondents to answer these types of questions during a telephone interview. Furthermore, not all of the attributes, such as residential environment and architectural style, have clear 'good' and 'bad' endpoints. For these reasons, the use of simple rating scales was chosen to elicit the values. A benefit of the current approach is that the resulting single- and multi-attribute utilities are measured using some type of global scale and not on a local scale (Monat 2009). The latter means that the poorest choice among the local options – in the above described example this concerns a range from 1 to 6 rooms – would obtain a score of 0, and the best 100. In the case of a global scale, the poorest choice among the entire universe of choices can get a score of 0, and the best 100. The drawback of a local scale is that it may over-emphasize the importance of small differences in values and may consequently lead to wrong decisions (Monat 2009).

Secondly, the decision problem was simplified. A number of attributes that probably influence the decision for a particular place of residence were not included, such as, for example, distance to work and school, distance to services and public transport and distance to the dwellings of relatives and friends. Furthermore, the attribute levels were also limited. For example, detached houses were not included in the research. The third potential limitation is that the multi-attribute utilities were calculated using the linear additive preference function. An important limitation of this approach is that it allows small advantages on some attributes to compensate for a large disadvantage on another. Thus, the possibility that thresholds of unacceptable performance exist for some criteria is neglected. Note, however, that Burnett (2008) showed in a study into shopping travel behaviors that compensatory rules (utility maximization) were quite important in a number of situations, especially for higher order goods (high involvement goods).

Finally, the application of Multi-Attribute Utility theory described in this chapter did not include trade-offs between attributes. Payne et al. (1999) propose that it is in making trade-offs that one's values are most often revealed to oneself and to outside observers. Timmermans et al. (1994) argue that when trade-offs are not included, the measurement task might not reflect the mechanisms underlying the actual decision-making and choice processes and may not be realistic to respondents. Multi-Attribute Utility theory in its purest form does indeed include trade-offs between attributes. Keeney and Raiffa (1976, p. 82 and following) describe methods to compose preference structures and value functions in the case of two or three attributes. The value functions are based on trade-offs. For example, suppose that the attribute size of the backyard has the value of 30 for a size of 5 m and the attribute number of rooms has the value of 80 for five rooms. With what amount should the size of the backyard increase in order to make up for the loss of one room? These kinds of questions are repeated for different levels and values of the attributes. However, such questions may not be easy to answer and the number of questions may become quite large if there are more than two or three attributes. Besides, respondents might not like having to make trade-offs because it is emotionally and cognitively burdening (Payne et al. 1999). Forcing respondents to make trade-offs might lead to behaviors such as selecting the status quo option, unwillingness to trade-off at all and delaying choice. For these reasons, trade-offs were not included in the study described in this chapter.

There are also some limitations that apply to Multi-Attribute Utility theory in general that have to be mentioned. Firstly, the theory presumes that rational decision-making takes place, i.e., more utility is preferred to less utility. The theory supposes that human values may only influence consumer choices by affecting what product attributes consumers prefer and that it is the calculated evaluation of product attributes that in turn determines product choice (Allen 2002). However, consumers also make emotionally laden, intuitive and holistic judgments about products. Allen (2002) suggests that human values shape consumers' product choices in two ways. Firstly, human values may influence the importance of the products' attributes, which in turn influences product preference. Secondly, human values may influence product preference directly by making an affective, intuitive and holistic judgment.

Secondly, the assumption that the importance of the attribute is independent of the level of the attribute may not hold as the importance of the attributes may be dependent upon the range of the scale over which the value function is defined. The weight could change when the range of the scale changes. For example, a respondent may be indifferent about the size of the backyard when the range is limited to between 5 and 15 m, because he does not consider a backyard interesting if it does not have a length of at least 25 m. Anything less is deemed unimportant. However, when the range stretches from 5 to 500 m, then backyard size does indeed become important to him. The solution is to choose a range of attribute levels that is wide enough to appeal to most respondents.

Related to the previously mentioned limitation is the third issue, which is that respondents may not be able to provide evaluations for a distinct attribute level without taking related attributes into account. The importance or value of a particular attribute may be dependent upon the level of other attributes. For example, the size of the living room may not be important to some respondents. However, it may become important when the number of rooms in the dwelling is very small. In our study, it seemed that the lower level of the attribute of purchase costs of € 140,000 was influenced by assumptions about the size or state of maintenance of such a relatively cheap dwelling. Timmermans et al. (1994) and Molin et al. (1996) wonder whether respondents are capable of expressing their evaluation of separate housing attributes, not knowing what to assume about the values of the remaining attributes influencing their preferences. As our respondents did not mention having problems providing values for individual attribute levels and the results showed face-validity for all attributes, except purchase costs, the problem may be manageable.

Fourthly, the multi-attribute approach does not allow the testing of the appropriateness of the chosen preference function (for example, additive or multiplicative) to combine the single-attribute utilities into an overall utility, unless some external criterion is available, such as an overall evaluation or overt behavior (Veldhuisen and Timmermans 1984).

Finally, the preferences obtained using Multi-Attribute Utility theory may not represent real housing choices. This is because housing choices may frequently reflect the dominance of constraints rather than preferences (Maclennan 1977). Note, however, that this drawback is not limited to Multi-Attribute Utility theory but applies to all methods based on stated preferences. Furthermore, it does not mean that housing preferences should not be examined. Housing preferences may be especially valuable in a time when the housing market becomes more and more demand-oriented, as we see nowadays.

In conclusion, despite the limitations, Multi-Attribute Utility theory may have additional value in the field of housing preferences research. It provides the possibility of examining the importance and attractiveness of separate dwelling characteristics, to calculate single-attribute utilities, to calculate overall utilities for combinations of attribute levels, to distinguish consumer groups with different preferences, and to choose amongst alternatives when different alternatives are available.

References

Allen, M. (2002). Human values and product symbolism: Do consumers form product preference by comparing the human values symbolized by a product to the human values that they endorse? *Journal of Applied Social Psychology, 32*, 2475–2501.

Barreiro-Hurlé, J., & Gómez-Limón, J. A. (2008). Reconsidering heterogeneity and aggregation issues in environmental valuation: A multi-attribute approach. *Environmental and Resource Economics, 40*, 551–570.

Bettman, J. R., Luce, M. F., & Payne, J. W. (2006). Constructive consumer choice processes. In S. Lichtenstein & P. Slovic (Eds.), *The construction of preference*. New York: Cambridge University Press.

Boumeester, H. J. F. M., Hoekstra, J. S. C. M., Meesters, J., & Coolen, H. C. C. H. (2005). *Woonwensen nader in kaart: de woonbeleving van bewoners*. Voorburg: NVB Vereniging voor ontwikkelaars en bouwondernemers.

Boumeester, H. J. F. M., Coolen, H. C. C. H., Dol, C. P., Goetgeluk, R. W., Jansen, S. J. T., Mariën, A. A. A., & Molin, E. (2008a). *Module Consumentengedrag WoON 2006, Hoofdrapport*. Delft: Onderzoeksinstituut OTB.

Boumeester, H. J. F. M., Mariën, A. A. A., Rietdijk, N., & Nuss, F. A. H. (2008b). *Huizenkopers in Profiel. Onderzoek naar wensen van potentiële huizenkopers*. Voorburg: NVB Vereniging voor ontwikkelaars en bouwondernemers.

Breij, I., de Hoog, R., & Zandvliet, L. (1989). Computer ondersteund onderzoek naar woon-voorkeuren. In S. Musterd (Ed.), *Methoden voor woning-en woonmilieubehoefte onderzoek*. Amsterdam: SISWO.

Burnett, P. (2008). Variable decision strategies, rational choice, and situation-related travel demand. *Environment and Planning A, 40*, 2259–2281.

Canbolat, Y. B., Chelst, K., & Garg, N. (2007). Combining decision tree and MAUT for selecting a country for a global manufacturing facility. *Omega, 35*, 312–325.

Edwards, W., & Newman, J. R. (1982). *Multiattribute evaluation*. Beverly Hills: Sage.

Floor, H., & van Kempen, R. (1994). Wonen op maat. In I. Smid & H. Priemus (Eds.), *Bewonerspreferenties: Richtsnoer voor Investeringen in Nieuwbouw en de Woningvoorraad* (pp. 13–32). Delft: Delftse Universitaire Pers.

Goetgeluk, R. (1997). Bomen over wonen, woningmarktonderzoek met beslissingsbomen (Dissertation, University of Utrecht, Utrecht: Utrecht Geographical Studies), p. 235.

Heins, S. (2002). Rural residential environments in city and countryside: Countryside images, demand for and supply of rural residential environments. (Dissertation, University of Utrecht, Delft: Uitgeverij Eburon).

Jansen, S., Boumeester, H., Coolen, H., Goetgeluk, R., & Molin, E. (2009). The impact of including images in a conjoint measurement task: Results of two small-scale studies. *Housing and the Built Environment, 24*(3), 271–297.

Jia, J., Fischer, G. W., & Dyer, J. S. (1998). Attribute weighting methods and decision quality in the presence of response error: A simulation study. *Journal of Behavioral Decision Making, 11*, 85–105.

Keeney, R. L., & Raiffa, H. (1976). *Decisions with multiple objectives: Preferences and value tradeoffs*. New York: Wiley.

Latinopoulos, D. (2008). Estimating the potential impacts of irrigation water pricing using multi-criteria decision making modelling. An application to Northern Greece. *Water Resource Management, 22*, 1761–1782.

Lindberg, E., Garling, T., & Montgomery, H. (1989). Belief-value structures as determinants of consumer-behavior – A study of housing preferences and choices. *Journal of Consumer Policy, 12*, 119–137.

Linkov, I., Satterstrom, F. K., Kiker, G., Batchelor, C., Bridges, T., & Ferguson, E. (2006). From comparative risk assessment to multi-criteria decision analysis and adaptive management: Recent developments and applications. *Environment International, 32*, 1072–1093.

Maclennan, D. (1977). Information, space and measurement of housing preferences and demand. *Scottish Journal of Political Economy, 24*, 97–115.

Meyer, V., Scheuer, S., & Haase, D. (2009). A multicriteria approach for flood risk mapping exemplified at the Mulde river, Germany. *National Hazards, 48*, 17–39.

Molin, E., Oppewal, H., & Timmermans, H. (1996). Predicting consumer response to new housing: A stated choice experiment. *Journal of Housing and the Built Environment, 11*(3), 197–311.

Monat, J. P. (2009). The benefits of global scaling in multi-criteria decision analysis. *Judgment and Decision Making, 4*(6), 492–508.

Park, W. C., Hughes, R. W., Thukral, V., & Friedmann, R. (1981). Consumers' decision plans and subsequent choice behavior. *Journal of Marketing, 45*(Spring), 33–47.

Payne, J. W., Bettman, J. R., & Johnson, E. J. (1993). *The adaptive decision maker*. Cambridge: Cambridge University Press.

Payne, J. W., Bettman, J. R., & Schkade, D. A. (1999). Measuring constructed preferences: Towards a building code. *Journal of Risk and Uncertainty, 19*, 243–270.

Raju, K. S., & Vasan, A. (2007). Multi attribute utility theory for irrigation system evaluation. *Water Resource Management, 21*, 717–728.

Timmermans, H., Molin, E., & van Noortwijk, L. (1994). Housing choice processes: Stated versus revealed modelling approaches. *Journal of Housing and the Built Environment, 9*, 215–227.

Veldhuisen, K. J., & Timmermans, H. J. P. (1984). Specification of individual residential utility functions: A comparative analysis of three measurement procedures. *Environment and Planning A, 16*, 1573–1582.

von Winterfeldt, D., & Edwards, W. (1986). *Decision analysis and behavioral research*. Cambridge: Cambridge University Press.

Vreeker, R. (2006). Evaluating effects of multiple land-use projects: A comparison of methods. *Journal of Housing and the Built Environment, 21*, 33–50.

Chapter 6
Conjoint Analysis

Eric J.E. Molin

6.1 Introduction

Conjoint analysis is in essence a data collection method that is especially useful if one is interested in examining the trade-offs that individuals or households make with respect to residential characteristics. The quintessence of this method is that (hypothetical) residential profiles are constructed that describe residential alternatives in characteristics that are assumed to influence residential preference or choice. The characteristics of the house and the residential environment are called attributes, of which housing type, and price are examples, and their values are called attribute levels, for example, single-family and multi-family house, and 500 and 1,000 Euros per month. In so called conjoint (or stated preference) experiments residential profiles describing different combinations of attribute levels are shown to respondents. Respondents are then requested to evaluate each residential profile on a rating-scale or to choose between two or more presented profiles. The observed responses are then decomposed into the part-worth utility each attribute level contributes to the overall utility respondents derive from residential alternatives. To that effect, a proper statistical analysis technique, such as regression or logit models, should be applied dependent on the measurement level of the observed responses. Hence, the result of a conjoint analysis is a utility function that describes to what extent each attribute contributes to the overall utility of residential alternatives.

As will become clear in this chapter, the estimated utility function may provide insight into a range of issues related to residential preference, of which the following are probably the most important. First, the estimated utility function indicates to what extent each attribute level contributes to the overall utility. For continuous attributes it may additionally provide insight into the form of the utility function, for example, one can test whether this relationship is linear or curvilinear. Second,

E.J.E. Molin (✉)
Faculty of Technology, Policy and Management,
Delft University of Technology, Delft, The Netherlands
e-mail: e.j.e.molin@tudelft.nl

S.J.T. Jansen et al. (eds.), *The Measurement and Analysis of Housing Preference and Choice*, DOI 10.1007/978-90-481-8894-9_6,

the utility function provides insight into the importance of each attribute, in other words which of the attributes, as varied in the experiment, has the largest impact on utility. Third, it provides insight into the trade-offs among the residential attributes. For example, as a costs attribute is included, the willingness to pay for improvements of other attributes can be derived. Fourth, if the conjoint experiment is properly constructed, it allows one to test for interaction-effects (see Sect 6.2 for an explanation). Fifth, the utility function enables one to predict the overall utility for any residential alternative expressed in the attributes that are varied in the conjoint experiment. Hence, this allows one to predict which of a series of alternative residences is preferred. Finally, if the conjoint experiment is structured as a choice task, the estimated choice model can predict how the choice probabilities are distributed across available alternatives within various scenarios.

At least one of these aspects has led researchers to apply conjoint analysis to model residential preferences for more than three decades. It has been applied to examine a wide range of issues related to residential preferences, such as to model the preferences of specific groups, such as students (Louviere and Henley 1977; Louviere 1979), divorcees (Timmermans and van Noortwijk 1995; Timmermans et al. 1996) and rental tenants (Quigley 1985; Walker et al. 2002); preference for specific areas, such as inner-city neighborhoods (Phipps and Carter 1984, 1985), and rural areas (Joseph et al. 1989; van Dam et al. 2002); transport-related characteristics (Borgers and Timmermans 1993; Borgers et al. 2008; Katoshevski and Timmermans 2001; Hunt et al. 1994; Tayyaran et al. 2003); the impact of environmental amenities (Earnhart 2001); choices in transitional housing systems (Donggen and Li 2004); complex decision-making involving many attributes (Timmermans 1989; Louviere and Timmermans 1990; van de Vyvere et al. 1998), and group decision-making (e.g. Timmermans et al. 1992; Borgers and Timmermans 1993; Molin 1999; Molin et al. 1997, 1999, 2000).

The aim of this chapter is to provide an introduction to the application of conjoint analysis to the modeling of residential preferences. To achieve this aim, the Sect 6.2 of this chapter introduces the utility function. The Sect 6.3 discusses the construction of the conjoint experiment. The Sect 6.4 provides an empirical illustration by reporting on a study examining the residential preferences of potential residents of a planned neighborhood. It should be noted that due to space limitations, this chapter mainly focuses on the basics of conjoint analysis. Section 6.5 and final section discusses the limitations of the basic methods and briefly discusses more advanced methods that overcome these limitations. For more details, the reader is referred to more exhaustive publications, such as Louviere 1988 and Louviere et al. 2000.

6.2 Utility Models

In order to properly construct the conjoint experiment, one should have a clear idea of the utility function one intends to estimate. Therefore, the utility function is discussed in this section, while the construction of the experiment is discussed in the next section.

6.2.1 *The Utility Function*

As already explained in the Introduction, the utility function describes to what extent the attributes influence the overall utility derived from residential alternatives. Typically it is assumed that utility consists of a structural part that can be explained by the estimated model and an error component, the part of utility than cannot be explained. Furthermore, it is typically assumed that the structural part of utility is a linear summation of part-worth (or marginal) utility contributions of the attributes, which can be expressed as follows:

$$U_j = V_j + \varepsilon_j = \beta_0 + \sum_{i=1}^{I} \beta_i X_{ij} + \varepsilon_j \tag{6.1}$$

where,

U_j = overall utility attached to alternative j;
V_j = the structural component of utility; that part of the utility that is determined by the model;
ε_j = an error component or random part of utility; that part of utility that is not determined by the model;
β_0 = the utility constant;
β_i = the coefficient to be estimated for attribute *i*;
X_{ij} = the value of attribute *i* describing alternative *j*;
$\beta_i X_{ij}$ = part-worth (or marginal) utility contribution of attribute *i* to the overall utility of alternative *j*.

6.2.2 *Effects Coding*

In order to include categorical attributes into the utility function, like, for example, tenure or dwelling type, the attribute levels need to be coded. Different coding schemes can be applied, such as dummy coding, effects coding, and orthogonal coding. These coding schemes differ in the direct interpretation of the estimated coefficients, but do not differ with respect to the resulting part-worth utilities of the attribute levels. As effects coding is applied in the example study presented later in this chapter, only this coding scheme is described in more detail here. An advantage of effects coding is that the estimated utility constant can be interpreted as the average utility attached to the residential alternatives included in the experiment. In the more well-known dummy coding scheme, the estimated constant denotes the utility of the alternative of which all levels are coded by zero. This is only relevant if that alternative serves as a benchmark alternative for which it is relevant to compare all other alternatives to. As this is often not the case, effects coding usually offers a more attractive interpretation.

Table 6.1 provides the coding scheme for effects coding for two, three and four level attributes. This scheme indicates that L levels are coded by L-1 indicator variables iv. For each indicator variable l, a coefficient ßl. is estimated. Hence, for

Table 6.1 Effects-coding for two-, three- and four-level attributes

	Indicator variables (iv)			Part-worth utility
Two levels	iv_1			
1	1			$ß_1$
2	−1			$-ß_1$
Parameters:	$ß_1$			
Three levels	iv_1	iv_2		
1	1	0		$ß_1$
2	0	1		$ß_2$
3	−1	−1		$-(ß_1+ß_2)$
Parameters:	$ß_1$	$ß_2$		
Four levels	iv_1	iv_2	iv_3	
1	1	0	0	$ß_1$
2	0	1	0	$ß_2$
3	0	0	1	$ß_3$
4	−1	−1	−1	$-(ß_1+ß_2+ß_3)$
Parameters:	$ß_1$	$ß_2$	$ß_3$	

a four-level attribute, three parameters ß1, ß2, and ß3 are estimated. The estimated coefficients are used to calculate the part-worth utility (pwu) of each level by multiplying the estimated parameter with its code and summing the results across the indicator variables (coded columns). The resulting part-worth utilities expressed in the estimated coefficient ß1 are presented in Table 6.1. From the table it becomes clear that the sum of the part-worth utilities across the levels of a particular attribute is, by definition, zero.

6.2.3 *Interpretation of Part-Worth Utilities*

The interpretation of the estimated part-worth utilities when effects coding is applied will be demonstrated by a simple example. Assume that in a conjoint experiment respondents evaluated eight residential profiles describing three attributes which each are varied in two levels. As respondents evaluated the profiles on a ten-point rating scale, a regression model was estimated that resulted in the part-worth utility contributions presented in Table 6.2. As effects-coding was applied, the estimated intercept or constant is equal to the mean overall utility derived from all the profiles included in the experiment. Hence, the mean utility derived from the residential profiles included in the experiment is equal to 5.

The part-worth utilities can be interpreted as the contribution of the attribute levels to the overall utility expressed as the deviation from the constant, thus from the mean overall utility. Hence, a positive part-worth utility means that the presence of the attribute level in a residential alternative increases the total utility derived from that alternative, and consequently, a negative part-worth utility decreases the overall utility.

Table 6.2 Estimated (hypothetical) part-worth utilities

Attributes and levels	Codes	Estimated coefficient	Part-worth utility
Constant			5
Tenure		0.5	
Owner-occupied	−1		0.5
Rental	1		−0.5
Housing type		−1	
Single-family house	−1		1
Multi-family house	1		−1
Monthly costs		−2	
500 euro	−1		2
1,000 euro	1		−2

The part-worth utilities presented in Table 6.2 indicate that owner-occupied houses increase utility, while rental houses decrease utility. Hence, owner-occupied houses are preferred to rental houses. Likewise, Table 6.2 indicates that single-family houses are preferred to multi-family houses and that cheap houses are preferred to expensive houses. Furthermore, considering the size of the estimated coefficients of the part-worth utilities, Table 6.2 indicates that of all the attributes, monthly costs has the largest impact on utility, followed by housing type and finally tenure.

Based on the utility function just discussed, one is able to predict the utility for any combination of attribute levels included in the experiment. For example, the utility for an owner-occupied single-family house of 500 euro per month is predicted as: 5 (constant) $+0.5+1+2=8.5$. Furthermore, by applying linear interpolation the utility contribution can be predicted for any value of a continuous attribute that falls within the range varied in the experiment. In this example, this would only apply to the attribute monthly costs.

6.2.4 *Interaction Effects*

The utility function just discussed is a main-effects-only model. A main-effect is the utility contribution of an attribute level to the overall utility irrespective of the presence of any other attribute level in the alternative. For example, our example model predicts that the combined effect of a rental multi-family house is: $-0.5-1=-1.5$ utility points. However, the part-worth utility contribution of an attribute level may not always be independent from other attribute levels in the alternative. This means that a specific combination of two attribute levels may have a different effect on utility than the sum of their associated part-worth utilities (the main-effects). This difference is denoted by an interaction-effect, which can be modeled by estimating a coefficient for the product of the attributes.

Hence, an interaction-effect can be regarded as a correction of the sum of the main-effects. For example, assume that the interaction-effect for tenure and housing type is equal to 0.25, this indicates that the joint utility of rental and multi-family house is 0.25 larger than the sum of their main effects (since the product of their codes is +1). Likewise, based on this interaction effect, the joint combination of single-family and owner-occupied house is 0.25 larger than predicted by the main-effects, while the joint combination of both rental - single-family house and owner-occupied – multi-family house is 0.25 smaller than predicted by the main-effects.

In addition to the two-way interactions just described, higher order interaction effects may also play a role. Hence, the sum of the main-effects may be corrected by the specific combination of three or more attribute levels. However, two-way interaction effects are not often estimated in practice and higher-order interaction-effects are only very rarely estimated as these are hard to interpret.

6.2.5 Estimation of Rating-Based Models

As already discussed in the introduction, at least two different types of responses can be requested from the respondent in conjoint experiments: ratings or choices. Ratings involve evaluating each residential profile separately and expressing the result as a number on some preference rating scale. Hence, the overall utility U_j for each residential alternative j is directly observed. As the rating observations are assumed to be of interval level measurement, the observed overall profile ratings make up the dependent variable in a regression analysis, and the independent variables are formed by the coded attribute levels. Hence, these data are typically analyzed by applying ordinary least squares regression analysis, which, for example, can be conducted in SPSS.

6.2.6 Estimation of Choice-Based Models

If the conjoint experiment is framed as a choice task then the observed responses indicate whether an alternative was chosen (normally coded 1), whereas the remaining alternatives (coded 0) were not. Hence, the nominal-level data cannot be analyzed by applying ordinary least square regression analysis, but require the application of an appropriate limited dependent analysis technique.

A further difference with rating tasks is that in choice tasks utility is not directly observed, only choices among alternatives. By assuming that respondents choose for the alternative with the highest utility, however, utility can be linked to the probability that an alternative will be chosen. This requires making additional assumptions about the random component of utility ε_j. Typically, it is assumed that the errors are independently, identically distributed extreme values (the distribution is also referred to as: type I extreme value, Gumbell or double-exponential), which results in the well-known multinomial logit (MNL) model (e.g., Ben-Akiva, and Lerman 1985; Train 2003):

$$p_j = \frac{e^{V_j}}{\sum_{j' \in S} e^{V_{j'}}} \tag{6.2}$$

Where p_j is the probability of choosing alternative j; S denotes the choice set of j alternatives; and e^{V_j} denotes the exponent of V_j, the structural part of the utility. MNL models are typically estimated in specialized software packages, such as Nlogit (Limdep), Alogit or Biogeme.

In order to provide a complete introduction to conjoint analysis, both choice-based and rating-based models will be covered in this chapter. However, it should be noted that most academics now prefer choice-based models (e.g. Louviere et al. 2000, 2010) for reasons that will be explained later (see subsection about measurement tasks).

6.3 Constructing the Conjoint Experiment

Once a utility function is assumed, the researcher can start constructing a conjoint experiment that allows the estimation of that function. This involves making decisions about: (1) selection of attributes, (2) determination of attribute levels, (3) choice of measurement task, (4) choice of experimental design, and (5) data collection. Each of these decisions is now considered in turn.

6.3.1 Selection of Attributes

The first step in constructing a conjoint experiment concerns the selection of the most salient attributes that influence the choice behavior of interest. This may be based on experience, literature research or preliminary research. Various small-scale qualitative research methods may be used for the latter, such as the repertory grid method, decision nets or tables, the factor listing method or focus group interviews (e.g., Molin 1999). On the one hand, this involves an attempt to select those attributes that influence the consumer choice behavior under investigation. On the other hand, it also involves a policy analysis to identify the "design" or "marketing" attributes that are relevant to policy-makers, urban designers, or marketers, even though these attributes may not necessarily be the most relevant to consumers.

To identify and define the most salient attributes one typically considers the following criteria (Louviere and Timmermans 1990): (1) which idiosyncratically attributes can be ignored in order to retain a list which is salient and relevant to most individuals most of the time; (2) which attributes can be retained, recombined, or re-expressed to keep the set of attributes as non-redundant and as small as possible to make an experiment tractable? (3) are the selected attributes relevant to managers or planners? That is, are they sufficiently specific to inform managers or planners about the consequences of policy or planning measures? and (4) are the attributes clearly defined to the respondents? That is, do the terms used connect with the respondents' cognitive representations of choice alternatives?

Typically, the attributes and their values are described in text only. However, specific dwelling characteristics, such as architectural style, may be difficult to describe in words. This may lead to the use of visual methods for such attributes, which may enhance the realism of the task, thus increasing the external validity of the resulting models (e.g. Dijkstra and Timmermans 1997; Orzechowski et al. 2005). However, images – especially photographs of existing dwellings – may also include elements that are not controlled for by the experimental design, such as color, the type of brick, weather conditions and additional objects as cars, garbage bins, and people (e.g. Jansen et al. 2009). These non-controlled elements may influence the respondents' choices, but their effects cannot be estimated due to insufficient variation across the alternatives, resulting in biased parameter estimates. Hence, the initial gain in validity may be counterbalanced by the negative impact of the non-controllable elements. Drawing realistic computer images in which the variation of all the visual elements is based on an experimental design may be a better alternative. However, as many aspects are related to housing design, one probably needs very large experimental designs to control for all these characteristics, which would probably result in too many profiles to handle in practice. Due to these drawbacks, there seems to be a consensus among researchers about not including images in conjoint experiments unless absolutely necessary due to a specific interest in measuring preferences with respect to design issues.

6.3.2 Determination of Attribute Levels

Once attributes have been selected one has to decide which levels to use to construct the profiles. For categorical attributes, the range of levels is often fixed by certain constraints. For example, the attribute tenure has only two levels: rental houses and owner-occupied houses. The number of categories may also be determined by the degree of detail required to support managerial decisions. The number of attribute levels is usually limited to between two and four. For instance, the attribute dwelling type is often defined in terms of following four levels: apartment, row-house, semi-detached, detached.

For continuous attributes, one has to decide on the number of levels and on the range of the attribute values. The number of levels depends on the assumptions one is willing to make about the relationship between the attribute values and the derived utility. If one assumes that the part-worth utility linearly increases or decreases with increasing attribute values, only two attribute values are required. If there is an interest in testing for deviations from linearity or one assumes that optimum or minimum levels exists, one needs to select at least three levels, as a minimum of three points are needed to draw a curve. Finally, if one assumes that utility increases with increasing attribute values, but that one is indifferent with respect to middle values, one needs to select four values, which enables S-curved utility relationships to be estimated.

The range of attribute levels is chosen such that they span the range observed in current or planned choice alternatives. This principle is based on the idea that while it is more valid to interpolate preference values for attribute levels within the range of selected values, the validity of the utility of extrapolated attribute values may be of concern. If, for example, the goal of the model is to predict housing preferences for a certain new neighborhood and one expects that the cheapest house will cost 400 Euros per month and the most expensive house will cost 1,600 Euros per month, than these two values are chosen as the end points of the price range. Depending on the assumptions one is willing to make on the form of the utility function, one could add a single middle value of 1,000 Euros per month to arrive at a three level attribute or add the values 900 and 1,200 to arrive at a four-level attribute.

6.3.3 Choice of Measurement Task

The choice of measurement task deals with the kind of response that is requested from the respondents. Three different measurement tasks are generally distinguished. In addition to the rating and choice tasks mentioned before, ranking tasks also exist. A ranking task involves respondents rank-ordering the set of profiles with respect to their overall preference. Usually each of the profiles is described on a separate index card. Respondents can spread these cards out on a table, compare them, make a preliminary ordering and then change the order until they are satisfied with the final ordering. In a rating task, respondents are required to express their strength of preference on some rating scale for each profile. For instance, respondents are requested to rate the degree of attractiveness of each profile on a scale which runs from 0 (extremely unattractive) to 10 (extremely attractive). Thus, whereas a ranking task only measures preference order, a rating task also measures relative preference distances between profiles. Finally, in a choice task respondents choose between two or more profiles. Hence, the construction of choice experiments requires one additional step, because the profiles have to be placed into choice sets.

In the early days of the development of conjoint analysis methods, ranking tasks were the most common measurement task. Later, evaluating profiles on a rating scale became the dominant response format. Rating tasks are often found to be less difficult than ranking tasks, because respondents only need to evaluate each profile in turn instead of comparing all profiles at the same time. This is certainly the case if the number of attributes and consequently the number of profiles to be evaluated increases. Moreover, because in a rating task the profiles are shown to respondents one at a time, this allows one to describe the profiles subsequently in a questionnaire with the result that mail-back surveys are easier to develop.

Currently, choice tasks are the dominant measurement task. It is widely acknowledged that making choices more resembles the behavior of decision-makers in the real markets than rating each profile and therefore choice tasks result in more valid models. Furthermore, choice tasks enable one to estimate choice models that allow the prediction of choice probabilities. In principle, this can also be done by rating-based models, however, these models require one to assume a decision rule that relates ratings to choices. An example of such a decision rule is that decision-makers always choose for the alternative with the highest predicted utility. However, decision rules are assumed and cannot be tested for their validity. Finally, choice tasks allow the estimation of additional terms in the utility function that provide additional insight into residential choice behavior, thus may result in more behaviorally rich models. For example, availability terms may be estimated, that indicate to what extent the utility of an alternative is influenced by the availability of a competing alternative in the choice set (see final section).

As ranking tasks are often found to be rather difficult and are hardly used in practice any more, ranking tasks are not further discussed in this chapter.

6.3.4 *Choice of Experimental Design*

In this section, experimental designs are discussed that are used to construct the residential profiles and the choice sets. First, the construction of profiles is discussed, which may be used for rating tasks as well as for choice tasks. This is followed by a discussion on the construction of choice sets.

6.3.4.1 Constructing Profiles

The number and composition of the residential profiles that are presented to respondents is determined by the chosen experimental design. Basically, three different types of experimental designs can be distinguished: full-factorial designs, fractional-factorial designs and compromise designs.

A full-factorial design involves making all the possible combinations of the selected attribute levels that can be made. This design type allows the estimation of all main effects and all possible interaction-effects. A disadvantage of the full-factorial design type is that the number of profiles rapidly increases with increasing numbers of attributes and attribute levels. For example, in the empirical illustration discussed later in this chapter, a single two-level attribute and eight three-level attributes are selected to describe the residential profiles. A full-factorial design in this case would involve $2 * 3^8 = 13{,}122$ combinations. It may be obvious that this number is too large to handle in practical research. Hence, full-factorial designs are only applied if the number of attributes and number of attribute levels is very limited. As this is not usually the case in residential preference research, fractional-factorial designs are typically applied in practice.

A fractional-factorial design is usually an orthogonal selection of the full-factorial design. Orthogonal means that the attributes are not correlated across all the profiles. A consequence of this is that all the combinations that can be made between the levels of two attributes will occur in the resulting residential profiles. For example, the attribute level 'detached house' will typically be paired an equal number of times with high prices as with low prices. The advantage of uncorrelated attributes is that for rating-based experiments the lowest number of observations is required to arrive at statistical significance for the estimated coefficients. For choice-based experiments orthogonal designs provide a good starting point, but efficient design strategies may be even more efficient (see final section).

Another desired property of experimental designs concerns attribute level balance. This involves that all the levels of an attribute appear an equal number of times across all the constructed profiles. If this was not the case, than the estimation of the coefficient for one level of an attribute would be based on more observations than the estimation of another level. Consequently, the coefficient estimated on more observations would have a higher probability of becoming statistically significant. Providing attribute level balance in practice typically results in choosing an equal number of levels for each attribute or choosing a mix of two- and four-level attributes.

The use of fractional-factorial designs brings back the number of constructed profiles to manageable proportions. For example, the smallest possible orthogonal fractional-factorial design for the application, to be discussed later in the chapter, results in the construction of 27 profiles, which is considerably less than the 13,122 combinations of the full-factorial design. This reduction in the number of profiles comes at a cost. Fractional-factorial designs do not allow the estimation of interaction-effects. Hence, to apply this type of design one must assume that none of the interaction-effects for the attributes plays a role in the residential preferences, in other words that their effects are equal to zero.

Basically, a fractional-factorial design is a matrix in which the numbers dictate which attribute levels one should combine to create profiles. Each attribute is assigned to a column that varies a certain quantity of numbers of which the series (0, 1), (0, 1, 2) and (0, 1, 2, 3) are the most common. Hence, a column containing the first series of numbers can be used to vary a second-level attribute, the second series a third-level attribute and the third series a fourth-level attribute. Each row of the matrix represents a profile. To illustrate this, the simplest possible fractional-factorial design for three selected attributes is depicted in the left part of Table 6.3. This design allows three attributes to be varied with two levels each. The two levels

Table 6.3 Applying an orthogonal fractional factorial design to construct profiles

				Tenure: 0→ owner-occupied 1→ rental	Housing type: 0→ single family house 1→ multi family house	Price: 0→ 500 1→ 1,000
Nr.	Design				Profiles	
1	1	0	0	Rental	Single family house	500
2	0	1	0	Owner-occupied	Multi family house	500
3	1	1	1	Rental	Multi family house	1,000
4	0	0	1	Owner-occupied	Single family house	1,000

of each attribute are then arbitrarily assigned to a level 0 or 1 as presented in the upper part of the table. This results in the four profiles that are provided in the lower right part of the table. Hence, this design requires the assumption that all interaction-effects are equal to zero and allows only three of the main-effects to be estimated. Note that this design is included here only for illustrative purposes. To make profiles of three attributes each varying in two levels it would probably be better to construct a full-factorial design resulting in eight profiles, which has the advantage that it allows the estimation of all the interaction effects.

In order to arrive at an orthogonal fractional-factorial design one can make use of published designs (e.g. Addelman 1962; Steenkamp 1985), or use specialized computer software. For example, SPSS provides a conjoint analysis module that supports the construction of such designs. Another example is the recently introduced program Ngene, which is probably the most advanced software package to support the construction of statistical designs. The number of profiles constructed in this way is at least equal to the number of indicator variables one needs to estimate. However, because of the attribute level balance requirement explained earlier, the number of profiles is usually larger.

A disadvantage of the smallest fractional-factorial designs most often used in practice is that not all of the main-effects are independent of all the interaction-effects. As a consequence, if interaction-effects that were assumed to be zero are not zero in reality and thus played a role in the decision-making process, some of the main-effects are confounded with these interaction-effects. Hence, the estimated main-effects may then be biased. This can be prevented by selecting those designs in which all the main-effects are independent of all the two-way interactions. However, this requires selecting larger designs that increase the number of required profiles.

One way to construct such designs is to create the foldover of the selected fractional-factorial design. A foldover design is the mirror image of the original design. For two-level attributes this means that a 0 entry in the design is replaced by a 1 in the foldover design, and a 1 by a 0. For a three-level attribute, the entry 0 is replaced by 2, the entry 1 remains 1 and the entry 2 is replaced by zero. The foldover design created in this way is then added to the original fractional-factorial design. Fractional-factorial designs combined with their foldover have the property that all main-effects are orthogonal to unobserved, two-way interactions. A drawback of this design strategy is that it doubles the number of profiles to be evaluated.

The third and final experimental design type distinguished here is the compromise design. This design allows estimating some selected interaction-effects. A compromise design consists of a main-effects design, combined with a second design that permits the estimation of selected interaction effects (e.g., Louviere et al. 2000). Although compromise designs still require the assumption that certain interactions are zero, they usually do not extend the number of profiles as much as applying the foldover strategy does. For a further introduction in constructing designs the interested reader is referred to Steenkamp (1985), which also includes several of the most applied fractional designs, and Louviere et al. (2000).

It should be clear that the possibility to deal with interaction-effects needs to be traded-off against the requirement to limit the number of profiles to be evaluated by each respondent to avoid information overload. Decisions regarding design size are typically based on the assessment of how many profiles a respondent can reliably complete in a given response situation. In practice, fractional-factorial designs that only allow the estimation of main-effects are therefore often chosen, because these designs provide the lowest number of profiles and therefore limit the complexity and expense of the data collection effort. Main-effects-only-models often predict preferences for new choice alternatives reasonably well and it has been observed that main effects typically account for most of the variance in the data (Louviere 1988, p. 40). Still, as stated before, one has to be aware of the fact that for some of these designs, main-effects are not independent of interaction-effects, which may result in biased estimates.

6.3.4.2 Constructing Choice Sets

As indicated before, the construction of choice tasks requires that profiles are placed into choice sets. Hence, the construction of choice experiments with two or more residential alternatives requires an additional step compared to the construction of profiles in the rating-based approach. The simplest strategy for the construction of choice sets is that first residential profiles are constructed as discussed before, and these are then placed into choice sets. The number of alternatives or profiles in each choice set is then determined and fixed and the profiles are placed in the choice sets by randomly drawing from the pool of profiles until all profiles are placed. This strategy leads to a minimum number of required choice sets, but does not ensure orthogonality between the profiles in the choice sets. Hence, correlations will exist among the attributes of different profiles across all the choice sets. This results in a less efficient design, that is, more respondents have to be selected in the sample to arrive at statistically significant coefficients.

The following design strategy preserves the orthogonality of the attributes both within and between the alternatives. By this design strategy one simultaneously constructs choice alternatives and choice sets from a statistical design. In this case, an L^{M*N} design is used, where L is the fixed number of attribute levels, M the number of alternatives in each choice set, and N is the number of attributes in each alternative. Each attribute of each alternative is therefore treated as a separate factor and assigned to a column in the experimental design. This design can be applied to create so-called labeled choice alternatives, which means that each alternative in each choice set has a label or name, like for example, "single-family house" and "multi-family house". This design strategy allows different attributes to be included for each labeled choice alternative and/or different numbers or ranges to be chosen to vary the attributes levels for the differently labeled choice alternatives.

To illustrate this, let us apply this design strategy to the case included in Table 6.2. This involves creating choice-sets of two labeled alternatives each, a single-family and a multi-family house, hence M=2. In this case, the first

Table 6.4 Applying the L^{M*N} design strategy to constructing choice sets

Experimental design				Choice set	Single-family alternative		Multi-family alternative	
					Tenure	Costs per month	Tenure	Costs per month
0	0	0	0	1	Owner-occupied	1,000	Owner-occupied	800
0	0	0	1	2	Owner-occupied	1,000	Owner-occupied	400
0	1	1	0	3	Owner-occupied	500	Rental	800
0	1	1	1	4	Owner-occupied	500	Rental	400
1	0	1	0	5	Rental	1,000	Rental	800
1	0	1	1	6	Rental	1,000	Rental	400
1	1	0	0	7	Rental	500	Owner-occupied	800
1	1	0	1	8	Rental	500	Owner-occupied	400

alternative is always the single-family house and the second alternative is always the multi-family house. Each alternative is described by two attributes: price and tenure, hence N=2. Finally, each attribute is described in two levels, hence, L=2. Tenure is varied for both alternatives in the levels rental and owner-occupied and price in the levels 500 and 1,000 Euros for the single family alternative and 400 and 800 for the multi-family alternative. Then, an orthogonal fraction of a $2^2 * 2$ (thus a 2^4) full-factorial design needs to be selected to construct the alternatives and the choice sets. The first two columns of this design vary tenure and price of the single-family house and the following two columns vary the same two attributes of the multi-family house. This is illustrated in Table 6.4.

Irrespective of the chosen design strategy, one can add a base alternative to each choice set, which is often labeled as 'none of these alternatives'. In residential choice applications the base alternative is often labeled as 'do not move'. Labeled as such, it can be used to estimate the 'none-choice' or 'opt-out' choice, which means that given the residential alternatives available on the (hypothetical) housing market, one chooses not to move house but to stay put in one's current house. Hence, it allows one to predict the probability that residential choice alternatives will be chosen compared to not moving at all. The base alternative is given a utility of zero by definition and thereby sets the origin of the utility scale.

6.3.5 Data Collection

The guidelines for data collection are basically the same as those for regular survey questionnaires. This especially applies to sampling techniques, which are therefore not discussed any further here. The only remark that needs to be made here is that typically a series of responses are collected for every respondent, which rapidly increases the number of observations. Hence, conjoint experiments on average require a smaller sample size than regular surveys. A general rule of thumb is that at least 30 observations are required for each profile or choice set in order to estimate reliable models. It should be noted that if a base alternative is included in the choice experiment and this alternative is often chosen, this requires sampling more respondents.

As in regular survey research, collecting data by means of interviewers may be preferred to using self-explanatory written questionnaires, as interviewers are able to explain the measurement task to the respondent. However, because it is much cheaper, the majority of conjoint tasks are collected by means of written questionnaires, either by paper and pen postal surveys or by means of the Internet. One difference with regular surveys is that conjoint experiments are less suitable for data collection by telephone. This is because it is practically impossible to grasp a full understanding of residential profiles if these are orally described in more than a few attributes, certainly if respondents are requested to make choices between two or more residential profiles.

With respect to the treatment of missing data, basically the same rules apply as for regular surveys. In principle, one can estimate a model for all valid observations, including those from respondents that have missing values. It is good practice to carefully examine the missing data patterns. If a respondent has many missing data, this may indicate that the respondent did not understand the conjoint experiment and the researcher may consider removing the remaining responses of that person from the data. If choice data are collected, the problem of non-traders may appear, that is, respondents who always choose the same choice alternative and are thus not willing to trade-off the varied attributes. The researcher then has to decide whether these responses are valid and therefore should be kept in the analysis, or whether these responses are not valid and should therefore be removed. This non-trader problem has only recently been given attention in the literature (e.g. Hess et al. 2008).

6.4 An Illustration

The aim of this section is to illustrate the insights conjoint analysis can provide with respect to residential preferences as discussed in the introduction. In particular, the use of choice-based models is illustrated by reporting on a study modeling the residential choice of potential tenants of Meerhoven. Meerhoven is a relatively new residential district in the city of Eindhoven and is a so-called Vinex location, one of the building locations the Dutch government appointed in the past decade as an instrument to realize its compact city policy. This policy aimed to concentrate new housing in large-scale compact residential districts near existing cities in an attempt to safeguard rural areas. For the city of Eindhoven this meant that new housing had to be built at a higher housing density than was considered normal in the region. Furthermore, housing costs were above average as most of the housing was to be built without financial aid from central government. Therefore, new tenants had to come from middle and higher income groups. At the time of the research (1996), the general picture of these income groups was that they were not really willing to move house because most people in these groups already lived in good dwellings;

if they were willing to move house, they were looking for spacious housing, which was probably not in line with the Vinex location intentions. Therefore, the local government wanted to have more insight into the housing choice process of the target group, especially with respect to higher densities. A conjoint choice experiment was therefore developed, commissioned by the Eindhoven municipality and five housing cooperations.

It should be noted that this study was conducted in 1996 and thus reflects the housing costs of that time. As the goal of this illustration is to show the possibilities of conjoint analysis, no attempts have been made to transform the cost values into comparable current values. Furthermore, it should be noted that this study was part of a larger study on residential preferences in the Eindhoven region. Based on their responses in the preceding large-scale survey questionnaire, a target group of potential tenants of Meerhoven was selected for this study and divided into four segments of housing type preference. This example study reports only on the segment that has an initial preference for living in a semi-detached house (see Molin et al. 1996 for a report on the row-house segment, and Molin 1996 for the original research report that includes all segments).

6.4.1 The Experiment

The first step in building a stated preference model is to identify the attributes of interest. In the present study, attributes were identified on the basis of previous research efforts (e.g., Timmermans 1989; Louviere and Timmermans 1990) and discussions with experts, more specifically with representatives of the commissioning parties. The following attributes were selected to vary the costs and functionality of semi-detached houses: Tenure, monthly costs, number of bedrooms, size of living room, and depth of backyard. However, depth of backyard can also be used to vary the housing density. Likewise, more centralized parking and realizing more high-rise buildings in the housing district can also be used for this purpose. The attribute size of green space was included to examine whether creating a large central park could compensate for increasing the density elsewhere in the housing district. Finally, shopping centre was included to examine to what extent tenants were willing to use shopping facilities in neighboring districts. All attributes were varied in three levels, which allows the continuous attributes to test whether utility is linearly related to changing attribute values. Note that the monthly costs presented applied for both rental and owner-occupied; in the questionnaire it was explained which costs are covered. The list of selected attributes and their levels is portrayed in Table 6.5.

In the present study, it was assumed that all interaction effects were equal to zero. Hence, the smallest orthogonal fractional-factorial design to construct the residential profiles was selected, involving the construction of 27 profiles. These profiles were then randomly placed in choice sets of three profiles each. The option "do not move" was added to each choice set as a base alternative. In order to avoid order effects, the placement of profiles in choice sets was randomized nine times.

Table 6.5 Selected attributes and their levels

Tenure	Size of living room	Buildings in neighborhood
Rental	30 m^2	Mainly high-rise
Owner-occupied	40 m^2	Mixed low-rise and high-rise
	50 m^2	Mainly low-rise
Monthly costs	Depth of backyard	Green space
Nlg. 900	10 m	Large central park
Nlg. 1,200	15 m	A few fairly large public gardens
Nlg. 1,400	20 m	More small public gardens
Number of bedrooms	Car park	Shopping centre
2	Central in neighborhood	Outside district
3	In the street	Central (one big)
4	On private property	Neighborhood (a few small)

Respondents were requested to choose the housing alternative in each choice set that they were most likely to move into. If none of the housing alternatives in the choice set were acceptable, they could choose the option "do not move", which served as a base option in each choice set. Respondents were requested to complete nine choice sets, which included all 27 profiles, presented in one of the nine different random orders.

6.4.1.1 Sample

Stated choice data were collected in early 1996. Respondents were selected from a previous survey, which was primarily intended to collect data on the housing situation and housing needs in the region. In this data file, those households were selected which indicated that they (i) were willing to move house within 5 years, (ii) preferred new housing or had no preference regarding housing age, (iii) were looking for housing in Eindhoven, (iv) were willing to spend at least Nlg 700 a month on housing, (v) preferred a semi-detached house, and (vi) agreed to participate in a choice experiment. A questionnaire containing the conjoint choice experiment was sent by mail to the selected households. A total of 154 respondents were contacted by mail, and 99 of them returned the form resulting in a response rate of 64%.

The household type of the response group was distributed as follows: 12.1% single. 47.3% couples and 40.7% couples with children. On average, the households consisted of 2.7 persons. The person who earned the most in the household had an average age of 36 years. The mean average net income was about 3,700 Dutch guilders.

As housing choice is often the result of a multi-person decision-making process, multi-person households were asked to complete the questionnaire together with all the household members who were involved in the housing choice. Previous research found evidence for the hypothesis that models based on group tasks better predict group housing choices than models based on tasks completed by individual group members (Molin et al. 1999). In 62% of the multi-person households, at least two persons, mostly husband and wife, completed the questionnaire together. Additionally, in about a quarter of the households with children age 14 or older, at least one child participated.

6.4.1.2 Model Estimation

The observed responses in the choice experiment were aggregated into frequencies that indicated how often the various alternatives included in any particular choice set were chosen. This set of frequencies served as the dependent variable in the MNL model. A main-effects-only model was assumed, so the problem was to find the part-worth utilities that, given these assumptions, best reproduced the observed choice probabilities.

To that effect, the econometric software package Nlogit was used to estimate a multinomial logit model. Model estimation is based on the principles of maximum likelihood estimation, which involves maximizing the log-likelihood function (Ben-Akiva and Lerman 1985). The commonly used goodness-of-fit model measure, which indicates how well the estimated model is able to reproduce the observed choices in the experiment, is based on a comparison of the log-likelihood of the estimated model (LB) with the log-likelihood of the null model (L0), the model in which all parameters are assumed to be zero. In the present study, LB is equal to −329.03 and L0 is equal to −553.30. By these log-likelihoods, McFadden's Rho Square (the likelihood ratio index = 1 − LB/L0) is equal to 0.41. Similar to explained variance in regression analysis, higher McFadden's Rho values indicate higher model fit, but this measure typically has lower values than the R^2 (though this depends on the level of aggregation). Taking this into account, it can be concluded that the estimated model reproduces the observed choices well.

6.4.1.3 Part-Worth Utilities

The attribute levels were effects-coded. As explained before, two indicator variables are estimated for each three-level effects-coded attribute, the values of which are equal to the part-worth utility of the first two levels of each attribute. Consequently, only two t-values are presented in Table 6.6.

Table 6.6 shows that the utility constant is equal to –0.87, expressing the mean utility derived from all residential profiles included in the choice experiment. As the 'do not move' alternative was, by definition, given a utility of zero, the negative constant indicates that on average the residential profiles included are considered less attractive than the current residence. To illustrate this further, this means that if the potential tenants have a choice between the 'average profile' and 'not moving house', then according to the estimated model 70.5%[1] of the tenants will choose 'not moving house'. This confirms the earlier discussed belief that the target group

[1] The probability of choosing the *average alternative* (utility = –0.87) above the *do not move* option (utility = 0), is predicted by the MNL model as: $p = e^{-0.87}/(e^{-0.87} + e^{0}) = 0.295$.

Table 6.6 Part-worth utilities and t-values

	Part-worth utility	t-value	Sig.	Attribute importance
Constant	–0.87	–9.359	*	
Tenure				19.0%
Rent	–0.68	9.180	*	
Owner-occupied	0.68			
Monthly costs				19.9%
Nlg. 900	0.70	8.958	*	
Nlg. 1,200	0.03	0.350		
Nlg. 1,500	–0.73			
Number of bedrooms				14.6%
2	–0.65	–7.654	*	
3	0.25	3.118	*	
4	0.40			
Size of living room				6.4%
30 m^2	–0.22	–2.490	*	
40 m^2	–0.02	–0.275		
50 m^2	0.24			
Depth of backyard				9.3%
10 m	–0.37	–4.305	*	
15 m	0.30	3.631	*	
20 m	0.07			
Car park				11.3%
Central in neighborhood	–0.46	–5.180	*	
On the street	0.35	3.912	*	
On private property	0.11			
Buildings in neighborhood				10.2%
Mainly low-rise	0.30	3.997	*	
Mixed low-rise and high-rise	0.13	1.665		
Mainly high-rise	–0.43			
Green space				1.8%
Large central park	0.06	0.765		
A few fairly large public gardens	–0.07	–0.914		
More small public gardens	0.01			
Shopping centre				7.4%
Outside district	–0.27	–3.348	*	
Central (one big)	0.26	3.278	*	
Neighborhood (a few small)	0.01			

* Absolute t-values>1.96 indicate statistical significance at 0.05 level and are marked by *

is quite satisfied with their current residence. Hence, quality housing needs to be provided in Meerhoven in order to tempt the potential tenants to move house.

As explained before, a part-worth utility indicates the contribution of an attribute level to the overall utility of a residential alternative. Let us now summarize the most important results based on the part-worth utilities:

1. Owner-occupied houses are preferred to rental houses, as may be expected in the segment of residents preferring the semi-detached housing type.

2. Residential utility decreases with increasing monthly costs from 900 to 1,500 Dutch guilders. As the second estimated coefficient is not statically significant, it can be concluded that this relationship is linear. Hence, an increase in the monthly cost has the same effect on utility across the whole utility cost range.
3. Utility increases with increasing number of bedrooms. However, this relationship is not linear. The increase in utility between two and three bedrooms is much higher (+0.90 utility points) than the utility increase from three to four bedrooms (+0.15 utility points).
4. Utility linearly increases with the increasing size of the living room. This indicates that within the varied range of 30–50 m^2, each m^2 increase of the living room increases utility to the same extent.
5. With respect to the depth of the backyard, the utility distribution clearly shows an optimal value: 15 m is considered the most attractive depth (+0.30 utility points). A depth of 20 m comes second in attractiveness (+0.07 utility points), while a depth of only 10 m is the least preferred (−0.37 utility points).
6. Car parking on the street is clearly preferred to parking at a central place in the neighborhood. Surprisingly, parking on the street is also preferred to parking on private property.
7. A neighborhood with mainly low-rise buildings is preferred, although the utility difference between a mixed low- and high-rise neighborhood is not very large (0.17 utility points difference), suggesting that some high-rise buildings in the neighborhood are tolerated. On the other hand, a neighborhood with mainly high-rise buildings is clearly disliked.
8. None of the part-worth utility levels of concentration of green space are statistically significant. This indicates that tenants in this segment are indifferent with respect to the way green space is spread across the neighborhood. An alternative explanation is that tenants have a clear preference for one of the levels, but that the preference is equally spread across the three levels, which therefore cancels out at the average level. The estimated MNL model is not able to distinguish between these two explanations.
9. Using shopping centers outside the district is clearly disliked above using shopping centers within the district. Furthermore, using a central district shopping center is preferred above several smaller neighborhood shopping centers.

6.4.1.4 Attribute Importance

To quantify the total impact an attribute has on residential utility the attribute's relative importance can be calculated by considering the attribute's utility ranges. An attribute's utility range is the difference between the highest and the lowest estimated part-worth utility of its levels. The utility ranges are summed across all attributes and the perceptual contribution of each attribute to this sum is calculated. However, it should be noted that attribute importance is conditional on the selected attribute levels. For example, if a smaller range of price levels was selected, say Nlg. 900–1,300, the utility range in part-worth utilities would probably have been lower and therefore also its estimated importance.

The resulting relative importances of the attributes are presented in the last column of Table 6.6. This table shows that monthly costs is the most important attribute, closely followed by tenure. This is followed by the housing attribute number of bedrooms. Next in importance are the attributes related to housing density: car park, type of buildings in the neighborhood, and depth of backyard. The attributes shopping centre, size of living room and green space have the least impact on housing choice. It is remarkable that size of living room has such a low impact. Probably the lowest level of 30 m^2 is already considered sufficiently large.

6.4.1.5 Willingness to Pay

Because monthly costs was included as an attribute, the willingness to pay for improvements in housing quality can be calculated. The idea underlying this calculation is to find out by how much monthly costs can be increased to compensate for a utility increase due to an improvement in another attribute in order to keep the overall utility at the same level. For this calculation, the utility change related to the varied costs range is considered first. Table 6.6 indicates that an increase in costs from Nlg. 900 to Nlg. 1,500 decreases utility from 0.70 to –0.73, thus by 1.43 utility points. Hence, each utility point is worth Nlg. 600/1.43 utility points = Nlg. 418. Willingness to pay for a quality improvement can now be calculated by multiplying this amount by the utility increase due to a change in attribute values. For example, an increase from 2 to 3 bedrooms increases the utility by 0.90 utility points, which results in an estimated willingness to pay off 0.90 * 418 = 377 Dutch guilders more per month. Likewise, it can be calculated that potential tenants are only willing to pay an increase of Nlg. 63 per month for a further increase from 3 to 4 bedrooms.

6.4.1.6 Prediction of Choice Probabilities

As argued before, an important advantage of choice-based conjoint models above rating-based models is that choice probabilities for new housing alternatives can be predicted directly based on the estimated model, thus without making any untestable assumptions. To illustrate this, latent choices for residences in Meerhoven are predicted under varying density scenarios. Under the assumption that no other residences are available at the same time, the percentage of households that will choose for any of two specified houses or none of either type is predicted. It should be noted that all scenarios discussed here are formulated for illustrative purposes only and thus do not reflect any intended policy of the parties involved.

First, the choice probabilities are predicted for a base scenario including two houses that reflect typical semi-detached houses found in the Eindhoven region in the study period. The first house is relatively small and cheap, whereas the second house is relatively large and expensive. The values of all other attributes are kept the same for both houses, with the exception of the type of environment: House 1 is located in a mixed low- and high-rise neighborhood, whereas house 2 is located in a mainly low-rise environment. The attribute values of both houses can be found in the top part of Table 6.7.

To examine the effects that increasing the housing density has on choice probability, the values of three attributes related to housing density are changed. A first change involves increasing the number of high-rise buildings in the neighborhood by one level. The first house is then assumed to be located in a mainly high-rise environment and the second house in a mixed low- and high rise environment. The second change involves decreasing the depth of the backyard by 5 m, arriving at a backyard depth of 10 m. The final change involves that parking is centralized, hence, parking in the street is no long allowed. In a final step, the possibility that the disutility due to increased housing density can be compensated by improving other housing attributes – increasing the living room by 10 m^2, realizing a central shopping center in the new housing district and decreasing the monthly costs by Nlg. 300 – is examined. Note that all changes are cumulative, which, for example, means that the second change assumes that the first change has already taken place.

Table 6.7 shows that under the base scenario, the smaller and cheaper house 1, is somewhat more popular than the larger and more expensive house 2. Using this

Table 6.7 Choice prediction for two residences under varying housing density scenarios

Changes	House 1	House 2	Do not move	Total
	Owner-occupied Three bedrooms 30 m^2 living room Nlg. 1,200 per month 15 m backyard Car park in street Mixed low- & high-rise Large public gardens Shopping outside district	Owner-occupied Four bedrooms 40 m^2 living room Nlg. 1,500 per month 15 m backyard Car park in street Mainly low-rise Large public gardens Shopping outside district		
Base	30.6%	25.1%	44.3%	100%
Increasing density[a]				
More high-rise	21.1%	25.5%	53.4%	100%
5 m smaller backyard	14.0%	16.9%	69.2%	100%
Central car park	7.5%	9.1%	83.4%	100%
Improving quality[a]				
+10 m^2 living room	9.1%	10.8%	80.0%	100%
+Central shopping center	13.6%	16.1%	70.2%	100%
−Nlg. 300 per month	20.3%	26.3%	53.5%	100%

[a]All changes are cumulative: for example, 5 m smaller backyard assumes more high-rise has already been applied

scenario, it is predicted that almost half of the potential tenants would not move to any of these two houses. The table further shows that the propensity not to move house rapidly increases with increasing density.

Table 6.7 shows that in the densest housing environment, 83.4% of the potential tenants is not inclined to move house, illustrating that potential tenants are quite sensitive to housing density. The question of whether the disutility due to the increased housing density can be compensated for is answered by considering the lowest three rows of Table 6.7. These results indicate that the choice probabilities do increase due to the proposed improvements, but that these improvements are not sufficient to fully compensate for the housing density effect.

It should be noted that the predicted choice probabilities should be interpreted with care. As already discussed, the predictions are based on the assumption that no other housing alternatives than those included in the scenarios are available for the prospective tenants. This is of course a very stringent assumption as housing alternatives are constantly being added to and removed from the housing market and at any moment in time it is likely that more than two alternatives will be available for most prospective occupants. A full simulation of the housing choice would require taking the dynamic aspect of the housing market is into account. Hence, the predictions should not be interpreted as absolute predictions of market shares of the residential alternative, but more valued for providing the possibility to compare the relative impact that possible scenarios have on changes in housing choice.

6.5 Advanced Methods

As indicated in the Introduction, this chapter introduced the most basic conjoint analysis methods for modeling residential preferences. These basic methods have their limitations, which may be overcome by applying more advanced methods. The most relevant of these methods for modeling residential preferences are briefly discussed in this section.

6.5.1 Hierarchical Information Integration Approach

Typically many attributes influence residential preference, however, there is a limit to the number of attributes that can be included in conjoint experiments. Many researchers assume that most respondents can handle up to seven or eight attributes at a time. However, there is much debate about this limit in the literature as others have shown that respondents can successfully handle up to 13 attributes (Louviere 1979). If the researcher wishes to include many attributes in the experiment, but at the same time wishes to limit the number of attributes in the residential profiles, the

Hierarchical Information Integration (HII) approach may provide an alternative (e.g. Louviere 1984; Timmermans 1989; Louviere and Timmermans 1990; Molin 1999). This approach assumes that if prospective tenants are confronted with complex decision problems involving many attributes, they first group the attributes into higher order decision constructs, for instance into a housing construct and a neighborhood construct. It is further assumed that tenants then make trade-offs among attributes that belong to a decision construct to form impressions of each of the decision constructs and therefore arrive at a housing evaluation and a neighborhood evaluation. Finally, it is assumed that they integrate these decision construct evaluations into the overall evaluation of the residence.

These assumptions are straightforwardly followed to construct HII conjoint experiments. A separate experiment is constructed for each decision construct, which varies only the attributes belonging to that construct. In addition, a so-called bridging experiment is constructed to examine the integration of the decision construct evaluations. If the attributes in the bridging experiment are expressed in terms of the levels of the rating scale used in the construct experiments to evaluate the construct profiles, then a single concatenated model can be constructed. Oppewal et al. (1994) proposed a variant on this method involving the construction of Integrated HII experiments, which has the advantage that it does not require constructing a separate bridging experiment. For more details about the HII approach, its variants and a review of HII applications, the interested reader is referred to Molin and Timmermans (2009).

6.5.2 *Efficient Design Strategies*

The design strategies discussed in this chapter are based on orthogonal designs to construct the residential profiles. A recent stream of publications, however, argues that orthogonal designs do not always result in the most efficient model estimation (e.g., Bliemer and Rose 2006; Rose and Bliemer 2006, 2009; Rose et al. 2008). This literature advocates constructing choice sets by using so-called prior estimates. Prior estimates are best guesses on the real coefficients, which may be based on small-scale preliminary research. These are then used to create choice sets in which the alternatives have about equal utility. This strategy avoids the occurrence of dominant alternatives, which are alternatives that have at least a higher part-worth utility on a single attribute, while they have no lower part-worth utility on any other attribute. As dominant alternatives will always be chosen by respondents, a choice set including a dominant alternative does not provide any information on the trade-offs among the attributes. As by using priors dominant alternatives are avoided and choice sets are created in which utility is more balanced, each observed choice reveals maximum information about the trade-offs. Hence, efficient designs potentially require a smaller number of respondents to arrive at statistical significance of the estimated coefficients than orthogonal designs do. The earlier discussed software Ngene supports the construction of efficient designs.

6.5.3 Modeling Heterogeneity

A limitation of the MNL model as presented in the empirical illustration is that this model does not take heterogeneity in tastes among the prospective tenants into account. Hence, the preferences of all tenants are represented by a single coefficient and possible preference differences among respondents, also called taste-heterogeneity, are only represented by the error term. However, taste-heterogeneity can be taken into account in conjoint models in at least three different ways: by estimating segment-based models, by estimating Mixed Logit (ML) models and by estimating Latent Class (LC) Models. Estimating segment-based models can be applied if choices as well as ratings are observed, while ML and LC models can only be estimated if choices are observed.

Segment-based models (e.g., Molin et al. 2001) involve estimating separate models for a-priori defined segments. For example, separate models may be estimated for the different categories of household type, which provides insight into the extent to which different household types have different residential preferences. Alternatively, one can include background variables in a single model by interacting these variables with the attribute coefficients, which allows a test to be conducted as to whether the coefficients are significantly related to the background variables included.

The more recently developed mixed logit (ML) model (e.g., Train 2003; Rouwendal and Meijer 2001) allows taking heterogeneity into account by estimating additional coefficients denoting the standard deviations for (selected) coefficients. This indicates to what extent individual tastes are distributed around an estimated mean parameter. For example, if an ML model had been applied in our illustration study, a standard deviation estimated for the attribute green space could have revealed whether tastes among the three levels really differed among the respondents or whether respondents were indifferent with respect to the way green space was distributed across the neighborhood. An advantage of the ML model is that it is extremely flexible and it is shown that this model can approximate any other discrete choice model.

Finally, Latent Class models assume that different homogeneous preference classes exist, which can be identified based on the observed choices (e.g. Boxall and Adamowicz 2002; Walker and Li 2007). The researcher determines the number of classes and a separate model is estimated for each of these classes together with a probability for every respondent of belonging to each of the classes. Compared to the Mixed Logit model, the Latent Class model is somewhat simpler to use and to interpret.

6.5.4 The IIA Assumption and Availability Effects

Although the MNL model is probably the most often applied choice model, it is often criticized for its stringent underlying assumptions and especially for its Independence from Irrelevant Alternatives (IIA) assumption. The IIA assumption implies that the utility of a choice alternative is independent from the presence of other alternatives in one's choice set and also independent from the attributes of the other alternatives. A consequence of this assumption is that the MNL model predicts that the choice probabilities of newly introduced choice alternatives draw to the same extent from choice probabilities all existing choice alternatives. This is illustrated with the following example. Imagine that at a certain moment in time, a potential tenant can choose between a multi-family house and a single-family house, both with a choice probability of 50%. Imagine also that an additional multi-family becomes available, with a choice probability of 30%. By the IIA property, the MNL model predicts that the probability of choosing both original houses drops to the same extent, in other words both drop by 15%. This, however, is highly unlikely: the newly introduced multi-family house has more in common with the original multi-family house and will probably also share attributes that are not included in the model, hence their error terms are likely to be correlated. It is therefore more likely that the choice probability of the single-family house will drop by a lower percentage than the predicted 15%, while the choice probability of the original multi-family house is likely to drop more than the predicted 15%. While this is what will probably happen in reality, the MNL model does not predict this due to its IIA assumption. Thus, the IIA assumption may pose problems in predicting correct choice probabilities for alternatives that are very similar, in other words that share characteristics that are not included in the model. On the other hand, if the utility function is properly defined and thus includes all attributes that play a role in residential choice, error terms are not correlated and the IIA assumption will not pose difficulties in predicting choices (Train 2003).

Whether or not the IIA assumption holds can be tested in various ways. One of these ways is by the estimation of the Universal or Mother Logit Model (McFadden 1975, see also McFadden et al. 1977). This model is a generalization of the MNL model in which the utility function is extended by so-called cross-effects. Cross-effects indicate to what extent the utility of an alternative increases or decreases by the availability of other alternatives in the choice set or by the presence of some of their characteristics. If none of these cross-effects are statistically significant, it can be concluded that the IIA assumption holds. The interested reader is further referred to Oppewal and Timmermans (1991), Timmermans and van Noortwijk (1995) and Timmermans et al. (1996), who discuss the design requirement for estimating cross-effects and illustrate the interpretation of the estimated cross-effects.

6.5.5 Epilogue

This chapter provides an introduction into the application of conjoint analysis to modeling residential preferences. The fundamentals of constructing conjoint experiments are discussed, and more advanced methods are briefly introduced. The method is illustrated using an empirical example in which attention was paid to the interpretation of the estimated model. It is hoped that this chapter contributes to the insights that conjoint analysis has to offer in examining residential preferences and that it will further stimulate the application of this method in this field of research.

References

Addelman, S. (1962). Orthogonal main-effects plans for asymmetrical factorial experiments. *Technometrics, 4*, 21–46.
Ben-Akiva, M., & Lerman, S. R. (1985). *Discrete choice analysis: Theory and application to travel demand.* Cambridge: MIT Press.
Bliemer, M. C. J., & Rose, J. M. (2006). Designing stated choice experiments: State-of-the-art. In *Proceedings of the 11th International Conference on Travel Behaviour Research.* Kyoto.
Borgers, A., & Timmermans, H. (1993). Transport facilities and residential choice behavior: A model of multi-person choice processes. *Papers in Regional Science, 72*, 45–61.
Borgers, A., Snellen, D., Poelman, J., & Timmermans, H. (2008). Preferences for car-restrained residential areas. *Journal of Urban Design, 13*, 257–267.
Boxall, P. C., & Adamowicz, W. L. (2002). Understanding heterogeneous preferences in random utility models: A latent class approach. *Environmental and Resource Economics, 23*, 421–446.
Dijkstra, J., & Timmermans, H. J. P. (1997). Exploring the possibilities of conjoint measurement as a decision-making tool for virtual wayfinding environments. In Y.-T. Liu (Ed.), *Proceedings of the second conference on computer aided architectural design research in Asia* (pp. 61–72). Taipeh: Hu's Publishers Inc.
Donggen, W., & Li, S. M. (2004). Housing preferences in a transitional housing system: The case of Beijing, China. *Environment and Planning A, 36*, 69–87.
Earnhart, D. (2001). Combining revealed and stated preference methods to value environmental amenities at residential locations. *Land Economics, 77*, 12–29.
Hess, S., Rose, J. M., Polak, J. W. (2008). Non-trading, lexicographic and inconsistent behavior in SP choice data. In *TRB: 87th Annual Meeting of the Transportation Research Board.* Washington, DC (Compendium of papers DVD).
Hunt, J. D., McMillan, J. D. P., & Abrahams, J. E. (1994). Stated preference investigation of influences on attractiveness of residential locations. *Transportation Research Records, 1466*, 79–87.
Jansen, S. J. T., Boumeester, H. J. F. M., Coolen, H. C. C. H., Goetgeluk, R. W., & Molin, E. J. E. (2009). The impact of including images in a conjoint measurement task: Evidence from two small-scale studies. *Journal of Housing and the Built Environment, 24*(3), 271–297.
Joseph, A. E., Smit, B., & McIlravey, G. P. (1989). Consumer preferences for rural residences: A conjoint analysis in Ontario, Canada. *Environment and Planning A, 21*, 47–65.
Katoshevski, R., & Timmermans, H. (2001). Using conjoint analysis to formulate user-centred guidelines for urban design: The example of new residential development in Israel. *Journal of Urban Design, 6*, 37–53.
Louviere, J. J. (1979). Modeling individual residential preferences: A totally disaggregate approach. *Transportation Research A, 13*, 373–384.
Louviere, J. J. (1984). Hierarchical information integration: A new method for the design and analysis of complex multiattribute judgement problems. In T. C. Kinnear (Ed.), *Advances in consumer research* (11, pp. 148–155). Provo: Association for Consumer Research.
Louviere, J. J. (1988). *Analyzing decision making: Metric conjoint analysis.* Sage University Paper, Series on Quantitative Applications in the Social Sciences (No. 07-067). Beverly Hills: Sage Publications.
Louviere, J. J., & Henley, D. H. (1977). An empirical analysis of student apartment selection decisions. *Geographical Analysis, 9*, 130–141.
Louviere, J. J., & Timmermans, H. J. P. (1990). Hierarchical information integration applied to residential choice behavior. *Geographical Analysis, 22*, 127–145.

Louviere, J. J., Hensher, D. A., & Swait, J. D. (2000). *Stated choice methods: Analysis and applications*. Cambridge: Cambridge University Press.
Louviere, J. J., Flynn, T. N., & Carson, R. T. (2010). Discrete choice experiments are not conjoint analysis. *Journal of Choice Modelling, 3*(3), 57–72.
McFadden, D. (1975). *On independence, structure and simultaneity in transportation demand analysis*. Working Paper 7511, urban travel demand forecasting project, Institute of Transportation Studies. Berkely: University of California.
McFadden, D., Train, K., & Tye, W. B. (1977). An application of diagnostic tests for the independence from irrelevant alternatives property of the multinomial logit model. *Transportation Research Record, 637*, 39–46.
Molin, E. (1996). Eindhoven doorgelicht. Deel 2: een model voor de woonkeuze in Meerhoven, Stichting Interface, Eindhoven (in Dutch).
Molin, E. J. E. (1999). *Conjoint modeling approaches for residential group preferences*. PhD. thesis, Faculteit Bouwkunde, Technische Universiteit Eindhoven, Bouwstenen 53.
Molin, E. J. E., & Timmermans, H. J. P. (2009). Hierarchical information integration experiments and integrated choice experiments. *Transport Reviews, 29*(5), 635–655.
Molin, E., Oppewal, H., & Timmermans, H. (1996). Modeling consumer response to new housing: A stated choice experiment. *Netherlands Journal of Housing and the Built Environment, 11*, 297–312.
Molin, E. J. E., Oppewal, H., & Timmermans, H. J. P. (1997). Modeling group preferences using a decompositional preference approach. *Group Decision and Negotiation, 6*(4), 339–350. Special issue on Innovative Approaches to Research.
Molin, E. J. E., Oppewal, H., & Timmermans, H. J. P. (1999). Group-based versus individual-based conjoint preference models of residential preferences: A comparative test. *Environment and Planning A, 31*(11), 1935–1947.
Molin, E. J. E., Oppewal, H., & Timmermans, H. J. P. (2000). A comparison of full profile and hierarchical information integration conjoint methods to modeling group preferences. *Marketing Letters, 11*(2), 165–175.
Molin, E. J. E., Oppewal, H., & Timmermans, H. J. P. (2001). Analyzing heterogeneity in conjoint estimates of residential preferences. *Journal of Housing and the Built Environment, 16*(3), 267–284.
Oppewal, H., & Timmermans, H. J. P. (1991). Context effects and decompositional choice modelling. *Papers in Regional Science, 70*, 113–131.
Oppewal, H., Louviere, J. J., & Timmermans, H. J. P. (1994). Modeling hierarchical conjoint processes with integrated choice experiments. *Journal of Marketing Research, 31*, 92–105.
Orzechowski, M. A., Arentze, T. A., Borgers, A. W. J., & Timmermans, H. J. P. (2005). Alternate methods of conjoint analysis for estimating housing preference functions: Effects of presentation style. *Journal of Housing and the Built Environment, 20*, 349–362.
Phipps, A. G., & Carter, J. E. (1984). An individual level analysis of the stress-resistance model of household mobility. *Geographical Analysis, 16*, 176–189.
Phipps, A. G., & Carter, J. E. (1985). Individual differences in the residential preferences of inner-city households. *Tijdschrift voor Economische en Sociale Geografie, 16*, 32–42.
Quigley, J. M. (1985). Consumer choice of dwelling, neighborhood and public services. *Regional Science and Urban Economics, 15*, 41–63.
Rose, J. M., & Bliemer, M. C. J. (2006, August, 16–20) *Designing efficient data for stated choice experiments: Accounting for socio-demographic and contextual effects in designing stated choice experiments*, Paper presented at the 11th International Conference on Travel Behaviour Research, Kyoto.
Rose, J. M., & Bliemer, M. C. J. (2009). Constructing efficient stated choice experimental designs. *Transport Reviews, 29*(5), 587–617.
Rose, J. M., Bliemer, M. C. J., Hensher, D. A., & Collins, A. T. (2008). Designing efficient stated choice experiments in the presence of reference alternatives. *Transportation Research B, 42*, 395–406.

Chapter 8
Lifestyle Method

Sylvia J.T. Jansen

8.1 Introduction

A classical approach in housing research is to predict housing demand on the basis of socio-demographic characteristics such as age, household status, and income. This method follows the idea that social background may both create opportunities and limit choices (Ganzeboom 1988). For example, persons with very low incomes usually do not live in large, detached houses and families with children usually prefer a single-family dwelling with a garden. A well-known model within this approach is the housing life cycle model (Rossi 1955). According to this model, changes in the size and composition of households and their preferences are related to different stages of nuclear family formation (cohabitation/marriage), expansion (birth of children), contraction (children moving out), and dissolution (death of a spouse). A transition into a new stage in the cycle may lead to a mismatch because housing characteristics, such as the number of bedrooms, might no longer meet the needs or preferences of the family.

However, people with the same background variables may have totally different preferences and behavioral patterns whereas people who score differently on the same background variables can share the same preferences and behavioral patterns (Michelson and Reed 1974; Wells 1974; Gunter and Furnham 1992; Pinkster and van Kempen 2002). Furthermore, demographic, socioeconomic and sociocultural shifts have taken place in Western economies in recent decades: households have become smaller and the variation in household types has increased. Other changes concern a greater variety of specific lifestyle-based subcultures and the expansion of the proportion of affluent households. These shifts have generated a broader variety in housing behavior (Kersloot and Kauko 2004).

Based on the trend of more differentiation, some argue that the traditional socio-demographic variables no longer suffice to explain and predict preferences

S.J.T. Jansen (✉)
OTB Research Institute for the Built Environment,
Delft University of Technology, Delft, The Netherlands
e-mail: s.j.t.jansen@tudelft.nl

S.J.T. Jansen et al. (eds.), *The Measurement and Analysis of Housing Preference and Choice*, DOI 10.1007/978-90-481-8894-9_8,

for the built environment and for developing local government in the area of housing (discussed in, for example, Driessen and Beereboom 1983; de Vreeze 1994; de Jong 1996; Reijndorp et al. 1997; Gibler and Nelson 2003; Heijs et al. 2005, 2009). Therefore, the motives underlying consumers' preferences are being explored. The choice of the dwelling can be understood as part of the person's general value orientation (Kersloot and Kauko 2004). Consumers usually act goal-oriented and choose a particular dwelling because they pursue values and goals that are important to them. For example, of three consumers with a similar income, one might buy a house because of the perceived social status of being a homeowner, one might buy a house because it seems like a good long-term financial investment, and the third might prefer a rental dwelling because of the perceived freedom of canceling a tenancy at any time. It is argued that by exploring underlying goals and values the preferences for dwelling (environment) characteristics can be better understood.

According to this approach, lifestyle variables are being proposed as an intermediary between the translation of socio-demographic characteristics into the determination of consumer preferences (Hustad and Pessemier 1974; Michelson and Reed 1974). Lifestyle is added to the traditional characteristics in the hope that it leads to more accurate explanations and predictions of consumers' preferences and choices. Proponents use such terms as to "put in the human factor," to enrich or to "put flesh on the bare statistical bones" of the segmentation based on socio-demographic variables alone. As Heijs et al. (2005, 2009) state, lifestyles would be able to fulfill this function, because they could close the gap between the traditional variables and the cultural aspects of life. Thus, lifestyle is used to improve the prediction of the housing demand in a quantitative as well as in a qualitative sense by obtaining more accurate measurements of consumers' preferences and choices. It is recognized, however, that socio-demographic and lifestyle variables may be related in different ways to housing preferences. Socio-demographic variables may determine what is attainable and what is needed and lifestyle variables (e.g., values or emotions) may determine taste. The type of housing – ground-plan, size, and cost – may be linked more to socio-demographic variables (income, age, size of household) whereas the appearance of the house may be particularly lifestyle-dependent (status, architecture, view, safety). For example, the household income and the number of persons in the household may determine the choice for a cheap dwelling with five rooms and values may determine the choice for a traditional architectural style.

Another application of lifestyle lies in the development of extraordinary housing projects, such as senior communities, an ecological neighborhood, dwellings with an architectural design based on the 1930s, a Mediterranean neighborhood, floating houses, gated communities, castle-like building blocks, and a neighborhood or small village entirely oriented toward water for recreational purposes. These extraordinary projects are directed at attracting consumers based on particular aspects of lifestyle that they share. The underlying motivation for such an approach can be that a developer wants to promote sales. However, these projects are also being developed out of idealistic considerations that some people have a need to live with like-minded people. Furthermore, an unusual location may ask for an exceptional approach. The housing research in these projects is focused on revealing the housing preferences of specific target groups, which are selected based on a specific aspect of lifestyle that they share.

Finally, another reason to perform lifestyle research is for management purposes, such as the allocation of (social) rented dwellings. Some argue that if residents with similar lifestyles lived in the same neighborhood, this could lead to more commitment to their own neighborhood and more mutual contacts and understanding. This concept is also proposed on a lower level, such as dedicating a block of flats with a shared entrance hall to residents having similar lifestyles. This would prevent conflicts and problems caused by deviating lifestyles. However, as the topic of this book concerns the measurement and analysis of housing preference and housing choice, the latter goal of applying lifestyle research will not be explored further here. So, note that the discussion provided in this chapter does not apply to the management, service, and allocation of dwellings and neighborhoods based on lifestyle.

8.2 Conceptual and Theoretical Aspects of Lifestyle

Lifestyle has been an area of interest to scholars in numerous disciplines, such as sociology, cultural anthropology, psychology, philosophy, marketing, and human geography. As a result, definitions of lifestyle differ between disciplines. However, the definition of lifestyle also differs within disciplines. Describing the similarities and differences between the various definitions of lifestyle is beyond the scope of this chapter. More information on the concept of lifestyle is provided in, for example, Michelson and Reed (1974), Wells (1974), Zablocki and Kanter (1976), Anderson and Golden (1984), Veal (2000), Ganzeboom (1988), Vyncke (2002), Pinkster and van Kempen (2002), van Diepen and Arnoldus (2003) and Heijs et al. (2005). Veal (2000) searched for a definition of lifestyle that is precise, unique, and efficient (i.e., including only those elements necessary for a precise definition). While developing this definition, he took account of the following topics: activities versus behavior, values and attitudes, individuals versus groups, group interaction, coherence, recognizability, and freedom of choice. I will not describe his search in detail but the wide range of topics he examined provides an indication of the difficulty of the concept of lifestyle. Ultimately, Veal (2000) proposes the following definition of lifestyle: "Lifestyle is the pattern of individual and social behavior characteristic of an individual or group." Note that a particular lifestyle is usually shared by a reasonable number of people, or else it would be idiosyncrasy or eccentricity (Chaney 1996, p. 11). Furthermore, lifestyle is usually expressed in behavior, but need not be.

8.2.1 Sociology

One of the first researchers who introduced the concept of lifestyle in 1922 was the sociologist Weber. He makes a distinction between class and status groups. Within classes, which are based on the way in which social groups are involved in the production process, Weber discerns status groups. These are groups of people who share the same standing and characterize themselves by expressing the same lifestyle through specific behavioral patterns. In this way, they can identify themselves with people of the same standing and differentiate themselves from groups with a different standing. The whole range of observable behaviors with this communicative and symbolic function is termed lifestyle by Weber (Pinkster and van Kempen 2002; Van Diepen and Arnoldus 2003).

Following this approach, some sociologists argue that lifestyles are a key factor in the definition of social status and can communicate social inequality (Felson 1976). Hamilton-Smith (2000) refers to Weberian theory by arguing that lifestyle is the summation of our response to the social world and portrays very clearly our place in the system of social prestige. Besides other factors, material consumption may offer a practical way of communicating social distinctions in everyday life. As an example of this field of research, a review of the relevance of consumer behavior, called material lifestyles, to social differentiation is provided by the sociologist Felson (1976). His review is directed at household consumer traits that can be found in living rooms, thus, that are visible to friends and acquaintances. The sociologists Zablocki and Kanter (1976) argued that lifestyles can no longer be determined by the economic system location alone but are also determined by factors outside the economic system. The authors assumed that the power of the economic and status systems in society was declining. They argued that when wealth and prestige markets fail to impose standards of taste and valuation in terms of relative position in these markets, individuals would seek other means of attaining value coherence. Thus, alternative lifestyles will not be formed at random, but rather in terms of shared strategies for regaining value coherence. See the Appendix for an overview of the various lifestyles as distinguished by Zablocki and Kanter.

Some sociologists involved residential preferences in their lifestyle research. Wirth (1938) established a theoretical model for exploring urbanism as a way of life. He argued that features of cities such as large numbers, high density and ethnic heterogeneity were related to the relative absence of intimate personal acquaintanceships, to anonymous, superficial, and transient relationships and to increased mobility and insecurity. Merton (1957) introduced the terms cosmopolitans and locals. Residents who are oriented to the world outside the local community were classified as cosmopolitans whereas those oriented toward the community were termed locals. Bell (1958) made a distinction between Familism (a high valuation on family living), Career (upward vertical mobility), and Consumership (striving for a high standard of living in the present). He observed that child and family orientations rather than career considerations were the prime motives for respondents with a middle-class status to move to the suburbs (described in Tallman and Morgner 1970). Furthermore, households with a lifestyle that is directed to making a career can usually be found in close proximity of urban facilities as these households want to spend their spare free time as efficiently as possible.

Another important researcher in the domain of lifestyle is the sociologist Pierre Bourdieu. He developed a theory of cultural capital and taste, which was published in France in 1979. Based on a large amount of data collected from about 30 different surveys, Bourdieu developed a two-dimensional model of lifestyle classification. One dimension or axis is formed by social-economic status, which is based on the total level of someone's resources: finance, level of education, standing of parents, and partner. The other dimension concerns the orientation of the resources: whether people derive their status from knowledge (especially knowledge about the cultural society) or from their financial background (assets, income, and capacities to earn money). Bourdieu shows that almost all lifestyles are two-dimensional; they are the result of the combination of someone's resources and the way in which they are employed. In the cultural hierarchy, one aims to demonstrate one's knowledge and cognitive capacities in the context of cultural society. Furthermore, purity and asceticism are emphasized. In the economic hierarchy, one aims to put on show income and properties, which was termed conspicuous consumption (Pinkster and van Kempen 2002).

8.2.2 Psychology

The psychologist Adler is also frequently mentioned as being the originator of the term lifestyle in 1929. Adler considered lifestyle to be the sum of the values, passions, knowledge, meaningful deeds, and eccentricities that constitute the uniqueness of each person. Lifestyle is some kind of guiding principle that each individual develops and which serves to organize and handle impressions of life. Lifestyle is thus considered to guide behavior. According to Adler, lifestyle is developed early in life in order to overcome some feeling of inferiority and to strive for superiority. However, Adler also believes that changes in behavior and character could take place when a person adopts new goals to strive for, indicating that lifestyle may change over time.

8.2.3 Marketing

In the domain of marketing, the lifestyle concept was introduced in the 1950s to understand, explain, and predict consumer behavior in order to focus marketing strategies (Anderson and Golden 1984). In 1963, Lazer introduced a concept of lifestyle that was based on three components (activities, interests, and opinions) (in Wedel and Kamakura 2000). Starting in the 1960s numerous lifestyle typologies were developed, as every product could have its own lifestyle typology. Typically, studies included up to 200 or 300 different items on activities, interests, and opinions. A data reduction technique, such as factor analysis, would then be used to obtain a smaller number of psychographic dimensions (Wedel and Kamakura 2000). In the 1970s a distinction was proposed between the term psychographics, for measures that are truly mental (attitudes, beliefs, opinions, personality traits, etc.), and lifestyle, for activities and behavioral reports (Wells 1974). However, this distinction does not have a widespread use as both lifestyle and psychographics are usually considered a combination of behavioral and psychological factors and are used interchangeably. Furthermore, some researchers argued that values, as the innermost drivers of behavior, would be more stable and generalizable than aspects such as activities and attitudes (Wedel and Kamakura 2000). Values were considered part of a value system. Such a system is defined as an enduring organization of values along a continuum of relative importance to the individual. The determination of value systems and the classification of consumers according to their value systems are likely to result in homogeneous groups in terms of the main motives underlying consumers' general behavior. Well-known instruments for measuring values and value systems are the Rokeach Value survey (Rokeach 1973), the List of Values (LOV, Kahle 1983), Values And Lifestyle Segmentation (VALS, Mitchell 1983) and the Schwartz Value list (Schwartz 1992).

8.2.4 Some Remarks Concerning the Definition of Lifestyle

Above, a number of theoretical approaches to lifestyle have been described. The approaches show important differences in their definition of lifestyle and in the factors through which it is expressed and through which it can be measured. The concept of lifestyle may vary from a limited characteristic to a broad spectrum of behavior and various psychological and social variables. Unfortunately, this problem is typical of the concept of lifestyle. In 1974, Michelson and Reed (p. 413) stated: "This review of the diverse definitions and applications of the notion of lifestyle makes clear, we think, the chaotic conceptual and operational state (and the consequent diminished usefulness) of lifestyle as a variable in social theory and research." Anderson and Golden (1984) also pointed this out by stating that: "Lifestyle is all things to all people, but this very fact that has made the concept appealing also impedes the development of further precision." However, despite the differences between the many definitions of lifestyle, the way in which most definitions agree is that the purpose of lifestyle is to provide a context within which the behavior of one or more actors can be understood, especially in terms of the stability, coherence, and purposefulness of action (Michelson and Reed 1974). Chaney (1996) argues that lifestyles help to make sense of (but not necessarily justify) what people do, why they do it, and what doing it means to them and to others.

8.3 Three Classification Approaches

In order to get a grip on the variation of characteristics embodied in lifestyle and to create order in the multitude and complexity of factors it is necessary to provide information on lifestyle in a manageable and comprehensible way. As Ganzeboom (1988) put it, the goal is to obtain a systematic classification that can be reproduced and that can be used to divide the population into distinct categories. Van der

Wouden and Kullberg (2002) argue that the construct of lifestyle loses its meaning if it does not yield any coherence within the preferences of individuals on various subjects. In general, there are three approaches to developing such systematic lifestyle typologies.

8.3.1 Less-Structured Data

The first approach is based on less-structured data that arise, for example, from observation, studying documents, and performing open interviews and focus group meetings. When an a priori category system or measurement scale is not available for collecting the data, the data are called less-structured (Coolen 2007). Such data have to be analyzed further and have to be categorized in order to be able to interpret the results. Thus, the inclusion level and the segmentation of the categories are not determined a priori, but they are constructed before, during, and after the collection of the data (Coolen 2007). An example of such a study is the study by Wirth (1938), who tried to establish a theoretical model to explore the urban way of living based on his observations, reflections, and the literature. The extraordinary projects that were mentioned in the Introduction are often performed using instruments that yield less-structured data, such as focus group meetings and in-depth interviews. The results are analyzed using traditional methods and are often presented only in reports and not in (peer-reviewed) papers as they are usually only of interest to the involved parties. Such research may boil down to "ordinary" housing market research, but only in specific target groups that are based on a shared aspect of lifestyle.

8.3.2 Coordinate System

The second approach uses a two- or three-dimensional coordinate system obtained with, for example, correspondence analysis or multidimensional scaling, on which alongside the axes the specific behaviors, orientations, and preferences are reflected. For example, the previously described sociologist Bourdieu positions lifestyle in a matrix with an axis for social-economic status from low to high status and an axis for the orientation of the resources (whether people derive their status from knowledge or from their financial background). Bourdieu shows that almost all lifestyles are two-dimensional; they are the result of the combination of someone's resources and the way in which they are applied. The transitions between different lifestyles are not clear-cut but move fluently. It is therefore difficult to pinpoint specific lifestyle groups with typical activities and preferences. For this reason, respondents are not classified in a specific number of lifestyle categories. This can be seen as a drawback of this approach. However, at the same time this approach is more realistic because in reality lifestyle categories do not show strict boundaries but instead have a fluent transition from one category to the other (Ganzeboom 1988).

8.3.3 Distinct Categories

The third approach uses distinct classification categories as obtained with methods such as cluster analysis and discriminant analysis. Usually, a benefit segmentation approach is followed as described by, for example, Gunter and Furnham (1992) and Wedel and Kamakura (2000). Firstly, data are collected on several variables that are related to the product concerned. Secondly, a clustering technique is used to divide the respondents into two or more (usually nonoverlapping) categories. Thirdly, clusters are generally described in terms of averages of the variables used in developing the clusters. Fourthly, the clusters are related to other variables that were not initially included, for example with the use of discriminant analysis. Fifthly, the resulting segments are examined for the use of advertising, target market definition, and so on. The number of discrete lifestyle categories differs among studies. However, most lifestyle typologies limit the number of classes to less than ten. A more detailed description of clustering methods is provided, for example, by Wedel and Kamakura (2000, Chap. 5). A drawback of this approach is that a large number of respondents may not fit into one of the distinct categories and may not be classified (Ganzeboom 1988). To solve that problem, overlapping and so-called fuzzy methods have been developed that allow cases to belong to more than one cluster at the same time (overlapping methods) or that provide information such as the degree of membership in each segment (fuzzy methods) (Wedel and Kamakura 2000).

8.3.4 Conclusion

In conclusion, three approaches are generally used for classifying people with similar characteristics: one based on less-structured data, one that uses a system of axes and one that divides people into distinct and separate categories. Note, however, that a combination of methods is also sometimes used, for example, in which segmentation groups are placed within a two-dimensional system.

8.4 A Brief Overview of Existing Lifestyle Typologies, Especially in the Domain of Housing

The Appendix provides an overview of existing lifestyle typologies, notably in the domain of housing. This table is not exhaustive, as typologies have been left out that were not developed specifically for use in the domain of housing. For example, in the domain of marketing numerous lifestyle typologies have been developed for every conceivable consumer good. Furthermore, typologies may also have been missed while studying the literature or new typologies might have emerged more recently. Note that some general typologies have been included, for example, those by Bourdieu and by Mitchell, because they have had such a large impact on lifestyle research in general.

The Appendix contains almost 40 different lifestyle typologies, of which more than 20 are specifically developed for the domain of housing. The Appendix shows that with the passing of time the number of lifestyle typologies has multiplied. Specifically, from 1999 to 2001, there was a boom in newly developed lifestyle typologies. Some of the typologies were developed by researchers (e.g., Bourdieu 1979), and others by commercial companies, such as SRI International (1989). Most lifestyle typologies are based on statistical techniques that result in a limited number of distinct categories (the third approach described above). The number of lifestyle categories usually does not exceed ten, and most have six to nine distinct categories. Most typologies that are specifically developed for use in the domain of housing have their roots in the Netherlands. It is not clear, however, whether this is the result of a specialization that only takes place in the Netherlands or whether it can be attributed to the relatively easy availability of these lifestyle typologies to the Dutch researcher.

8.5 Housing Preference Research Using Lifestyle Variables

After having provided insight into theoretical and practical aspects concerning the development of lifestyle typologies, the focus will now move to the practical use of lifestyle in the domain of housing. Previously, an important reason for the development and application of lifestyle typologies in the domain of housing was described: to support socio-demographic variables in predicting and explaining quantitative and qualitative housing demand. In the literature, a small number of studies was found that combined socio-demographic and lifestyle variables in order to explore housing preferences and choices.

Driessen and Beereboom (1983) composed seven lifestyle groups based on questions on recreational activities, such as doing sports, and activities with an obligatory character (work-related), such as doing physically strenuous activities during work. They explored whether these seven groups differed in their preferences with regard to 21 aspects of the urban residential environment, such as a quiet neighborhood, the availability of a garden and the safety of the neighborhood in the evening. The authors found differences between lifestyle groups for only 3 of the 21 aspects, namely: neighborhood with residents with the same social position, living 10 min from the center of the city and a respectable neighborhood. The authors conclude that the results using lifestyle groups are no better than the results based on lifecycle situation, social-economic position, and housing location alone.

Coolen et al. (2002) explored the added worth of values in the relationship between intended tenure choice and socio-demographic variables. The researchers showed that intended tenure choice could be predicted by age, current tenure, income, and household composition and that values have a small, but statistically significant, effect. The more important the values of power and achievement were deemed to be, the greater the tendency toward owning. Also, the more important family values were, the higher the tendency to rent.

Pinkster and van Kempen (2002) explored the relationship between socioeconomic and socio-demographic variables, lifestyle variables, and the preference for type of residential environment. They showed that household composition, age, income, education, and ethnicity could predict the preferred type of residential environment. Furthermore, they found significant relationships between preferred type of residential environment and lifestyle variables such as visiting museums, frequenting churches, and being interested in politics. In all cases, the amount of explained variance of the logistic regressions was larger if lifestyle variables were included than in the case when only socioeconomic and socio-demographic variables were included. Thus, lifestyle variables did have an added value and, furthermore, this influence was independent of the socioeconomic and socio-demographic variables included in the analyses.

Studied the worth of lifestyle variables in predicting a number of housing preferences, after correction for socio-demographic variables. The latter explained 0–45% of the variance in preferences. The lifestyle variables added between 0% and 5% of additional variance.

Based on the results of the studies described above, it seems that there are some small indications that lifestyle variables are able to predict specific aspects of housing preference or choice. However, one should not expect too great an impact from these variables.

8.6 An Example

Whereas the topic of this book is to provide an insight into methods and techniques for measuring housing preferences and choice, an example is now described to show one of the techniques that can be used to develop lifestyle typologies. Note that there are many techniques available, such as acquiring and analyzing less-structured (qualitative) data, contingency tables, log-linear models, regression analysis, discriminant analysis, cluster analysis, CHAID analysis, and so on. For a description see, for example, Wedel and Kamakura (2000). In the current example a multidimensional scaling (MDS) method is used, which provides a two-dimensional coordinate system (an example of the second approach mentioned in Sect. 8.3).

8.6.1 The Study

The data that are used in this example are collected in the context of a large study termed "Huizenkopers in Profiel," which has been performed every 2 years in the Netherlands since 1995. In this study, data on housing preferences and the current housing situation are obtained from respondents who have at least an average income, approximately 70% of all Dutch households. This criterion increases the probability that respondents actually have a choice with regard to carrying out their housing preferences in practice. The goal of the Huizenkopers in Profiel study is to determine the needs and wishes of future homebuyers in order to establish what has to be built. The data were collected though telephone interviews in the year 2010 from about 2,000 respondents. The analyses were performed only for respondents who indicated that they were willing to move if they could find a house that would fulfill all their needs, about 45% of respondents.

8.6.2 The Lifestyle Variables: Values

Lifestyle typologies can be based on numerous types of variables, such as opinions, activities, values, preferences, and personality characteristics. In the current example, lifestyle classification analyses were performed on values only. In particular, the data were collected using 29 values from the Schwartz Value List (Schwartz 1992). It is customary, for example in marketing, to base a lifestyle classification on values only. It is assumed that values are the innermost drivers of behavior, and, therefore, are reasonably stable and generalizable. Schwartz (1992) employs the following definition: values are (1) concepts or beliefs that (2) pertain to desirable end states or behaviors that (3) transcend specific situations, (4) guide selection or evaluation of behavior and events and that (5) are ordered by relative importance. Thus, values are viewed as the criteria that people use to select and justify actions and to evaluate people (including the self) and events (Schwartz 1992). Schwartz discerns 11 universal, motivational value types. These are provided here with some examples of values belonging to each type in parentheses: (1) Self-direction (freedom, creativity), (2) Stimulation (an exciting life, a varied life), (3) Hedonism (pleasure, enjoying life), (4) Achievement (ambitious, influential), (5) Power (social power, wealth), (6) Security (national security, reciprocation of favors), (7) Conformity (obedient, self-discipline), (8) Tradition (respect for tradition, devout), (9) Spirituality (a spiritual life, meaning in life), (10) Benevolence (helpful, responsible), and (11) Universalism (equality, unity with nature). It is assumed that every individual strives for values belonging to these domains. The composition of the various values one strives for is called a value system (Rokeach 1973). It is assumed that such value systems are relatively stable in time in the sense that they on average will be composed of the same values. However, the relative importance of each of the domains may change over time.

In the Huizenkopers in Profiel study of 2010, only 29 of Schwartz's original 56 items were included. This was because the questions regarding the values were part of a larger survey, as explained above. If too many questions were included, the interview burden for the respondents would be too high (fatigue, boredom, no more time, and so on). We slightly rephrased some of the values in order to make them more appropriate as a guiding principle in housing. As in the original survey by Schwartz (1992), each item was followed in parentheses by a short explanatory phrase, e.g., Family security (safety for loved ones). The questions were stated as follows: "How important do you find [the specific value] as a guiding principle in housing?" The respondent could provide an answer on a seven-point scale ranging from "not important at all" (1) to "extremely important" (7).

8.6.3 Multidimensional Scaling

Multidimensional scaling (MDS) is a technique used to uncover any structure or pattern that may be present in the data and to identify the dimensions on which subjects make their similarity judgments (Everitt and Dunn 2001). Notice that MDS concerns a class of techniques; it is not just one method. MDS attempts to find the structure in a set of proximity measures between objects, a so-called proximity matrix. This matrix can be based on direct measurements in which respondents are asked to assess the similarity of pairs of stimuli or, indirectly, as a measure of the correlation or covariance of a pair of stimuli based on the raw profile data (Everitt and Dunn 2001). The process of finding structure in the data is accomplished by assigning observations to specific locations in a low-dimensional space in such a way that the distances between points in the space match the given similarities or dissimilarities in the data as closely as possible. The result is a representation of the objects in that low-dimensional space, which will help to find structure in the data. In general, the larger the perceived dissimilarity between two data points, the further apart their points are placed in the resulting graphical model. The differences between data points are usually measured according to Euclidean distance. This technique is explained in, for example, Everitt and Dunn (2001), p. 93.

The method used in the present example is MDS (Proxscal: Proximity Scaling) based on Euclidean distance and standardized scores and with interval level proximity transformations.

8.6.4 Proposed Theoretical Structure of Values

Schwartz (1992) argues that if a sample of items adequately represents all aspects of a content domain, then those items will fill into the geometrical space quite evenly, with no major gaps and no distinct clusters. It is therefore necessary to partition the space into meaningful regions based on a priori theory of the conceptual notions among the values. In the current example, the obvious theoretical principle is the value structure as posed by Schwartz (1992). This structure is presented in Fig. 8.1. Note that the value type Spirituality is not represented in Fig. 8.1, as this value type was deemed somewhat unreliable by Schwartz. It is postulated by Schwartz (1992) that adjacent value types are the most compatible. Increasing distance indicates decreasing compatibility and greater conflict between values.

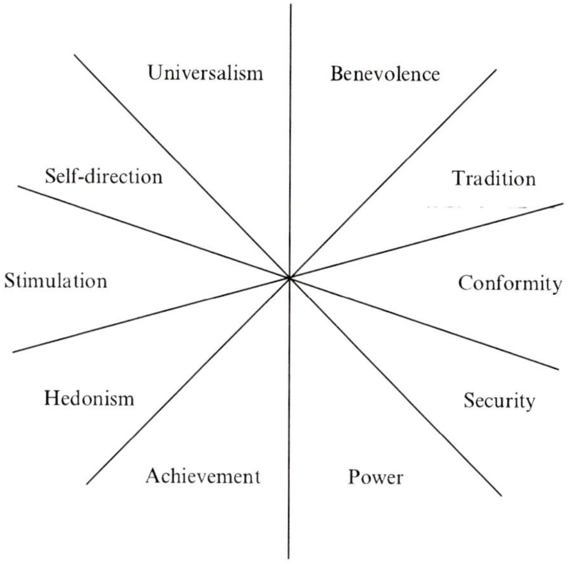

Fig. 8.1 Proposed theoretical structure of values by Schwartz (1992)

After performing the MDS analysis, a graphical two-dimensional solution is obtained with all the values located in the geometrical space according to their similarities (or dissimilarities). The fit of the solution can be determined by inspecting the goodness of fit criterion, called Stress. This criterion provides an indication of the differences between the points that represent the data and the observed proximities. In the current analysis, the stress is 0.066, which is judged to be in between fair and good (Kruskal 1964). In interpreting the results of the MDS analysis, the procedure described by Schwartz (1992) is followed. This means that the configuration of substantively related points that emerge to form regions and the arrangement of these regions in space relative to each other is interpreted. As a result, some values at the edge of a region may correlate less with other values of the same region than they do with some values on the edge of other regions. To indicate the regions, partition lines are drawn. These lines may be straight or curved, as long as they yield regions having continuous boundaries that do not intersect with the boundaries of other regions (Schwartz 1992). Following Schwartz, boundary lines were drawn by first connecting the values at the outer lines of the regions, avoiding any overlap between these boundaries. Next, the partition lines were placed between the boundaries. The results are provided in Fig. 8.2. Note that seven values were omitted from the analyses: we did not include values belonging to the Tradition value type scale because the mutual coherence between its values was too low (number of values=3, Cronbach's Alpha=0.51). Furthermore, we did not include the values "self-indulgence" and "sense of belonging," because they did not fit well into their expected scale based on previous

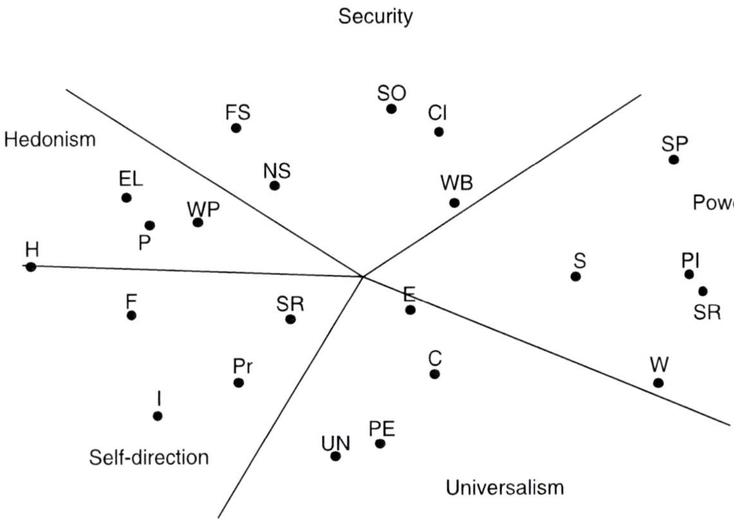

Fig. 8.2 Multidimensional scaling of 22 values. Note: *P* pleasure, *EL* enjoying life, *F* freedom, *C* creativity, *I* independent, *SR* self-respect, *Pr* privacy, *SP* social power, *W* wealth, *PI* preserving my public image, *SR* social recognition, *S* successful, *NS* national security, *FS* family security, *SO* social order, *H* healthy, *Cl* clean, *E* equality, *UN* unity with nature, *WB* a world of beauty, *WP* a dwelling environment at peace, *PE* protecting the environment

analyses using Cronbach's Alpha (see Jansen and Coolen 2010). Finally, two values were not included because they each represented one domain consisting of only one value.

From Fig. 8.2 it can be seen that the results generally conform to the theory posed by Schwartz (1992). There are some problems with four values (18%), which are not located within their hypothesized value type: a World of Beauty (WB; Universalism), Creativity (C; Self-direction), Healthy (H; Security), and a dwelling environment at Peace (WP; Universalism). However, this is quite customary. Schwartz proposed three criteria to decide whether a set of value points form a bounded region: (1) at least 60% of the values assigned to a particular value type must be located within this value type, (2) at most 33% of the values in a particular value type region may belong to any other value type, and (3) at least 70% of all values in a particular value type region must reflect the goal of the appropriate value type. In our analysis, all value types meet these requirements.

Note that an exact agreement between the current example and the results produced by Schwartz (1992) cannot be expected for several reasons. Firstly, only 22 of the original 56 values were included, leading to the omission of six value domains. Furthermore, a somewhat different phrasing of the question was applied as well as different answering categories (seven-point scale instead of a nine-point scale).

Table 8.1 Means and standard deviations of the value types

Scale	Number of values	Range of scores	*n*	Mean (Std)	Cronbach's Alpha
Security	5	5–35	893	31.69 (3.28)	0.73
Self-direction	5	5–35	892	30.74 (3.56)	0.72
Universalism	5	5–35	891	29.51 (3.96)	0.69
Power/achievement	5	5–35	892	20.58 (6.21)	0.77
Hedonism	2	2–14	908	13.03 (1.45)	0.70

8.6.5 Empirical Results

Now that the value structure has been revealed, indices of value types can be constructed in order to link values to their antecedents (e.g., socio-demographic variables) and consequences (e.g., behavior). The simplest index is to calculate, for every person, the mean importance of each value type based on the set of values that constitutes that type. This way, every respondent has an index score for each value type. The mean scores for the value types are presented in Table 8.1.

Note that, although four values were not placed within their matching scales according to the MDS analyses, further analyses using Cronbach's Alpha showed that their mutual coherence was good enough to form a scale (all scales had a Cronbach's Alpha of at least 0.69). Note further that respondents are not placed within one particular lifestyle category, such as "City-edge dwellers" but obtain a score for each of the five value types. These scores provide an indication of the importance of the particular value type for the respondent as a guiding principle in housing. Table 8.1 shows that, in general, Hedonism is believed to be the most important value type as a guiding principle in housing. This value type holds the values Enjoying life and Pleasure. The least importance is attached to the Power/Achievement value type, which contains values such as Social Recognition and Preserving my public image.

After calculating the value type scales, various analyses can be performed. As an example, the influence of value types on housing preferences was examined with the use of logistic regression analysis. After correction for the influence of socio-demographic variables, a number of relationships between value types and housing preferences were observed. Respondents who had indicated that they were willing to move if they could find a house that would fulfill all their needs were asked to provide their preferences with regard to a number of dwelling characteristics. Respondents who had indicated not to be willing to move were supposed to be satisfied with their current dwelling characteristics and these were taken as reflecting their preferences. The results are summarized in Table 8.2. Note that socio-demographic and lifestyle variables are included only if they show a statistically significant relationship to housing preferences.

Firstly, respondents who live in an owner-occupied dwelling find the value type Power and achievement less important as a guiding principle in housing than those who live in a rental dwelling, after correcting for age, education, household type,

Table 8.2 Relationships between value types and housing preferences

	R^2
Actual tenure: Rental (0) versus owner-occupied (1) ($n=540$)	28%
Age, education, household type, income	27%
Power and achievement: OR = 0.58	1%
Actual type of neighborhood: Quiet (0) versus lively (1) ($n=624$)	12%
Age, education	10%
Universalism: OR = 0.39	2%
Hedonism: OR = 5.41	
Actual residential environment: City (edge) (0) versus outside city (1) ($n=662$)	2%
Gender	1%
Universalism: OR = 1.85	1%
Preferred residential environment: City (edge) (0) versus outside city (1) ($n=816$)	4%
Household type	2%
Power and achievement: OR = 0.60	2%

R^2 percentage of explained variance, *OR* odds ratio

and income. The central goal of the value type Power is seen by Schwartz (1992) and Schwartz and Boehnke (2004) as attainment of social status and prestige, and control or dominance over people and resources. Both power and achievement values focus on social esteem. Intuitively, one should expect power values to be related to the preference for an owner-occupied dwelling instead of a rental dwelling. Moreover, Gibler and Nelson (2003) describe two studies in which ownership of a single-family detached house with a yard represents independence and success. In the Netherlands, however, having an owner-occupied home may not increase feelings of control or dominance. A recent study by Kleinhans and Elsinga (2010) showed that buying their home did not have an effect on Dutch homebuyers' feelings of control (i.e., influence of powerful others and of accidental happenings on life), self-esteem (i.e., having good personal qualities) and housing-related empowerment (i.e., being in control of their own home). Instead, renters may believe to have more control over their homes as they can cancel a tenancy at any time.

Secondly, respondents who live in a lively neighborhood find Hedonism more important and Universalism less important as a guiding principle in housing than respondents who prefer a quiet neighborhood, after correction for age and education. The motivational goal of Hedonism can be defined as pleasure or sensuous gratification for oneself (Schwartz 1992; Schwartz and Boehnke 2004). Lively neighborhoods might accommodate more facilities (e.g., shops, wellness centers, restaurants) that provide pleasurable experiences than quiet neighborhoods. The motivational goal of Universalism is understanding, appreciation, tolerance, and protection of welfare of all people and for nature (Schwartz 1992). Further analyses showed that the effect of Universalism could be largely attributed to the value of Unity with nature. Thus, respondents that live in a lively neighborhood find Unity with nature less important.

Thirdly, respondents who live in the city find Universalism as a guiding principle in housing less important than those who live outside the city, after correction for gender. Again, this effect can largely be attributed to the value Unity with nature. Finally, respondents that prefer to live in the city find Power and achievement more important as a guiding principle in housing than those who prefer to live outside the city, after correction for household type. This effect can largely be attributed to the value Social recognition. Thus, residents who prefer to live in the city might expect to find more social recognition there.

8.7 The Problems with Lifestyle

Despite the potential added value of lifestyle in housing research, the concept is wrought with problems and uncertainties. This section provides a description of the most important ones.

8.7.1 No Consensus About the Definition of Lifestyle

As noted previously, lifestyle is a complicated construct because there is no consensus in the research field about what is meant by lifestyle or what the factors are through which lifestyle is expressed. Such factors can be values, activities, personality traits, interests, and so on. Almost every lifestyle typology has made different choices with regard to the variables that make up the particular lifestyle categories. There has been little attempt to standardize items or concepts, with the result that the relation between the various existing typologies is unknown. Even worse, developers of lifestyle typologies do not make clear which variables constitute particular lifestyles, which variables determine lifestyles (antecedents), and which variables are the consequences of lifestyles (Heijs et al. 2005).

8.7.2 A Multitude of Lifestyle Typologies

The heterogeneity in lifestyle can also be seen in the multitude of lifestyle typologies, as shown in the Appendix. However, the simultaneous existence of so many lifestyle definitions and typologies need not necessarily be a problem, as every product or service may have its own suitable lifestyle typology. A lifestyle typology may focus only on users of particular products, such as journals or dwellings. This is called domain-specific segmentation. Lifestyle groups in one context may not be transposed to another context, as preferences in one domain might not be related to preferences and behavior in another domain (Pinkster and van Kempen 2002). Furthermore, typologies that are more general may be too crude to be applied in various circumstances. Research should focus on identifying psychological and behavioral factors that are relevant to the specific consumption behavior being investigated, rather than searching for some generic set of attitude functions. From this line of reasoning it follows that lifestyle research may be useful for local development or extraordinary housing projects for groups that are selected on a shared aspect of lifestyle, such as seniors and environmentally friendly consumers.

8.7.3 Information on Psychometric Properties Is Usually Lacking

For many lifestyle typologies neither detailed information on the development of the classification nor information on the psychometric characteristics, such as reliability and validity, is available. Many lifestyle typologies do not fulfill the criteria posed in scientific research for the coverage of research design and data analysis. This kind of information may be trade secret (Grünfeld 2003; Heijs et al. 2005), but it is also possible that extensive research into the psychometric properties of the lifestyle classification methods has not been performed yet.

8.7.4 The Translation of Lifestyle Variables into Dwelling Characteristics

An important question concerns the relationship between lifestyle and housing. How can lifestyle be translated into architectural and urban development plans? How do we know what has to be built, or rebuilt? How can, for example, a tendency to be oriented toward oneself be translated into physical dwelling characteristics? The relationship between housing preferences and lifestyle is complicated. Housing includes many aspects, for example physical aspects, such as the dwelling type and the number of rooms, and locational and social aspects, such as the number and distance of various facilities and the type of residents in the neighborhood. The gap between theory and practice is still wide with respect to the implementation of lifestyle variables in determining what has to be built.

8.7.5 The Stability of Lifestyle

Related to the points discussed here is the notion of the stability of lifestyles. Stability of lifestyles is a desirable property because housing is related to long-term investments and the housing market is inflexible, both in terms of responding to shifts in demand for specific types of housing and to shifts in the spatial concentration of the demand (Hooimeijer and Schutjens 1991). In this sense, frequently changing lifestyles are of little use (van der Wouden and Kullberg 2002). Dwellings and public spaces are usually built to last for a period of

50–100 years, whereas lifestyles may change much more frequently. It is generally assumed that changes in lifestyle usually occur gradually but lifestyles can also change rapidly, for example, related to sudden changes in family life such as a divorce or the birth of a child (Pinkster and van Kempen 2002). Other causes of lifestyle change may be fashion trends and increasing prosperity. Luxury goods, which used to be beyond the reach of people with low incomes, might now be affordable for most people.

8.8 Summary and Conclusion

This chapter provided insight into the development and use of the lifestyle concept in the domain of housing. Despite the prosperous growth of lifestyle typologies, the concept of lifestyle is still plagued by various troubles. There is also the question of whether it is really necessary to add lifestyle variables to the traditional socio-demographic variables. Research has not shown that traditional variables are inadequate or that lifestyle variables have an additional value in explaining or predicting housing preferences. A more useful application of lifestyle may be for local development and the development of extraordinary housing projects based on a shared aspect of lifestyle, such as senior communities or housing in a water-rich environment. The question, however, is whether this should be termed lifestyle research or boils down to traditional housing preference research in selected target groups.

Appendix 1 Overview of Lifestyle Typologies, Especially with Regard to Housing

Author(s)	Domain	Name	Year	Lifestyle categories
Bastiaansen (1997)	Housing	Brabant ongemonteerd	1997	Villager, landscape dweller, yuppie, anarchist, traditional, principled, pleasure-lover, hypermobile, attached to one place, in lodgings, idealist[a]
Bell (1958)	General		1958	Familism (a high valuation on family living), career (upward vertical mobility), consumership (striving for a high standard of living in the present)
De Bock and Decker (2009, Internet reference)	Interior		Around 2000	Design-lovers, party people, modern people, status people, sensitive people, respectable people[a]
Bourdieu (1979)	General	La distinction	1979	No specific categories. Two dimensions: (1) social-economic status and (2) the orientation of the resources
van Diepen and Musterd (2001)	Housing		2001	Four urban lifestyles: omni-urban lifestyle, economically necessary lifestyle, socially necessary lifestyle, non-necessary lifestyle[a]
van Diepen and Musterd (2009)	Housing		2009	Five urban lifestyles: superurban households, economically urban households, socially urban households, nonurban households, retired households
Driessen and Beereboom (1983)	Housing		1983	Active, non-active, less active, youth, workers, practically and usefully oriented, vital-expansively oriented[a]
van Engelsdorp Gastelaars (1980)	Housing		1980	Reserved community members, unreserved community members, reserved city-dwellers, unreserved city-dwellers[a]
van der Flier Consultancy (2009, internet reference)	Housing		1998	Original urbanites, original villagers, new urbanites, new villagers, edge-city dwellers, suburbanites, immigrants
Fournier et al. (1992)	General	Nine consumption lifestyles	1992	Functionalists, nurturers, aspirers, experientials, succeeders, moral minority, golden years, sustainers, subsisters
Gans (1968)	Housing		1968	Cosmopolitans, singles and childless, ethnic villagers, deprived, trapped and downward mobile
Ganzeboom (1990)	Housing	Empirical example	1990	No specific categories. Three dimensions: (1) economical hierarchy, (2) cultural hierarchy, and (3) stage of life dimension.

Gemeente Groningen, dienst RO/EZ (van der Schaaf 2002)	Housing	Groninger methode	Around 2000	Students, singles (young employed; not much economically active), couples (double income no kids; not much economically active; traditional households; active older people), older people (active older people; older people focused on their surroundings), families with children (not much economically active with children; traditional families, double income with kids), single-parent families[a]
Haest (1989)	Housing		1989	The old guard, different sort, foreigners[a]
Inbo and I&O Research (2009, internet reference)	Housing		2001	Constructor, neighbor, traditional, partner, unattached, hypermobile, dynamic individualist, quiet luxury[a]
Inbo (2009)	Housing	Soulife lifestyles	Around 2009?	Individualist, intractable, hypermobile, consumerist, world citizen, opportunist, quiet enjoyer, traditional, principled, comfort-seeker, landscape dweller, villager[a]
Merton (1957)	General		1957	Locals, cosmopolitans
Mitchell (1983)	General	Values and lifestyle segmentation (VALS)	1978	Need-driven: survivors, sustainers. Outer-directed: belongers, emulators, achievers Inner-directed: I-am-me's, experientials, societally conscious. integrateds: self-integrated
Motivaction Marketing Organization (Nijhuis and Schoemaker 2002)	Housing	Segmentatie model	1999	Domestic types, community thinkers, down-to-earth family clan, active individualists, busy middle classes, tolerant socializers, settled idealists
Motivaction Marketing Organization (Van der Lelij 2002)	General	Mentality	2002	Postmodern hedonists, post materialists, upwardly mobile, cosmopolitans, new conservatives, modern middle class, traditional middle class, leisure group
NFO-Trendbox (2009, Internet reference)	General	Value box	1995	Goal-focused adventurers, sober philosophers, spiritual altruists, order and respectability, the center, uncomplicated pleasure-lover[a]
National Readership Survey (2009, internet reference)	Housing	Super profiles classification	2005	Affluent achievers, thriving grays, settled suburbans, nest builders, urban ventures, country life, senior citizens, producers, hard-pressed families, have-nots, unclassifiable
De Nieuwe Unie (Overboom 2003)	Housing		2003?	Survivors, Discoverers, neighborhood focused, dynamic individualists, well-to-do families, modal seniors, well-to-do seniors[a]

(continued)

Appendix 1 (continued)

Author(s)	Domain	Name	Year	Lifestyle categories
Reijndorp et al. (1997)	Housing		1997	Uncertain existence: house as operating base, house as mobile home; rootless existence: Pied-à-terre, guesthouse; organized existence: combination household, network household; intractable existence: self-built paradise, do-it-yourself; shared existence: income neighborhoods, undivided city; recreational existence: outdoor city, holiday destination[a]
Reijndorp et al. (1998)	Housing		1998	Dwellers in new estates: outer-district dwellers, original urbanites, new urbanites, villagers, suburbanites, nomads[a]
de Rooij and Wallagh (2000)	Housing		2000	Urban professionals, hypermobiles, consumentists, active adults, city ecologists
Stuurgroep Experimenten Volkshuisvesting (2009, internet reference)	Housing	Woningatlas Consumentgericht Bouwen	1999	Villager, landscape dweller, yuppie, anarchist, traditionalist, principled, hypermobile[a]
Shevky and Bell (1955)	General		1955	Careerism, familism and consumerism
The SmartAgent Company (Kolpron Consultants and MarketResponse 2000)	Housing		2000	Dynamic individualists, free thinkers, community thinkers, entrenched, retiring group, quiet luxury
The SmartAgent Company (Hagen 2006)	General	Brand Strategy Research (BSR) model	2004	Red world, blue world, green world, yellow world
Sobel (1981)	General	Lifestyle and social structure	1981	No specific categories? Relationships between consumptive behavior and socio-demographic variables
SRI International (1989)	General	VALS 2	1989	Actualizers, strugglers, fulfilleds, achievers, experiencers, believers, strivers, makers
TNS NIPO (Hessing and Reuling 2003)	General	Waarden-segmenten in Nederland (WIN) model	Around 2003	Hedonists, business group, broad-minded group, commited group, caring group, conservatives, balanced group, luxury-seekers

Trendbox BV (in: Van Kralingen 2003)	General		1995	Modest social involvement, socially justified pleasure, concerned post-materialism, no-nonsense/dependent security, cultivated hedonism, self-conscious individualism, socially aware, religious altruism[a]
Urhahn Urban Design (1999)	Housing		1999	City-dweller, networker, dreamer, outdoors type, villager, do-er, comfort-based dweller, economically focused dweller, ostentatious dweller[a]
Value Group Ltd (2009, internet reference)	General	Insight value group scale	?	Self-actualizers, innovators, esteem-seekers, strivers, contented conformers, traditionalists, disconnected
Weber (1922)	General		1922	Divisions in society arising from class and status. Only illustrative examples of status groups are provided, no exhaustive categories
De Wijs-Mulkens (1999)	Housing		1999	No specific categories?
De Wijs-Mulkens and Ostendorf (2001)	Housing		2001	No specific categories?
Young and Rubicam (2009, internet reference)	General	Cross-cultural-consumer characterization (4Cs model)	?	Reformer, explorer, succeeder, aspirer, mainstreamer, struggler, resigned
Zablocki and Kanter (1976)	General		1976	Lifestyles determined by the economic system: property-dominated lifestyles, occupation-dominated lifestyles, and poverty-dominated lifestyles. Alternative lifestyles: regressive (hedonistic) lifestyles, etherealization lifestyles (spiritual, intellectual, monastic, and emotional lifestyles), community-based (ethnic lifestyle and communitarians), collective behavior–based lifestyles

Note: This table is not exhaustive as typologies may have been missed while studying the literature or new typologies might have emerged since. In general, typologies have been left out that were not developed specifically for use in the domain of housing. However, some general typologies have been included, because they have had such a large impact on lifestyle research in general. Finally, the table may contain flaws as it is very difficult to obtain the correct information. It just provides an indication of existing lifestyle typologies, especially with regard to housing.

[a]Translated from Dutch

References

Anderson, W. T., & Golden, L. L. (1984). Lifestyle and psychographics: A critical review and recommendation. In T. C. Kinnear (Ed.), *Advances in consumer research* (pp. 405–411). Provo: Association for Consumer Research.

Bastiaansen, I. (1997). Leefstijlen als basis voor het nieuwe wonen. In J. Broess & C. Grijzen (Eds.), *Brabant ongemonteerd* (pp. 5–20). Den Bosch: Projectburo Brabant 2050.

Bell, W. (1958). Social choice, life styles and suburban residence. In W. F. Dobriner (Ed.), *The suburban community*. New York: Putnam's.

Bourdieu, P. (1979). *La distinction; Critique sociale du jugement*. Paris: Editions de Minuit.

Chaney, D. (1996). *Lifestyles*. London: Routledge.

Coolen, H. (2007). Measurement and analysis of less structured data in housing research. *Open House International, 32*(3), 55–65.

Coolen, H., Boelhouwer, P., & van Driel, K. (2002). Values and goals as determinants of intended tenure choice. *Journal of Housing and the Built Environment, 17*, 215–236.

De Jong, F. (1996). *Woonvoorkeurenonderzoek: Theorie, empirie en relevantie voor de praktijk. Woonconsument en woningkwaliteit 5*. Delft: Uitgave Publicatiebureau Bouwkunde.

de Rooij, L., & Wallagh, G. (2000). *Wie Woont er aan de Zuidas? Leefstijlen en Woonarrangementen voor de Zuidas*. Amsterdam: Blauwhoed/Eurowoningen, MAB, Woonstichting De Key/De Principaal.

de Vreeze, N. (1994). Architectuur en bewonersvoorkeuren. In I. Smid & H. Priemus (Eds.), *Bewonerspreferenties: Richtsnoer voor investeringen in nieuwbouw en de woningvoorraad* (pp. 57–63). Delft: Delftse Universitaire Pers.

de Wijs-Mulkens, E. (1999). *Wonen op stand. Lifestyles en landschappen van de culturele en economische elite*. Amsterdam: het Spinhuis.

de Wijs-Mulkens, E., & Ostendorf, W. (2001). De smaak van het WBO. Smaak in twee dimensies. *Tijdschrift voor de Volkshuisvesting, 8*, 45–50.

Driessen, F. M. H. M., & Beereboom, H. J. A. (1983). *De kwaliteit van het stedelijk leefmilieu: Bewoners en hun voorkeuren*. Utrecht: Rijksuniversiteit Utrecht, Vakgroep Theorie en Methodologie van de Sociologie.

Everitt, B. S., & Dunn, G. D. (2001). *Applied multivariate data analysis* (2nd ed.). London: Arnold.

Felson, M. (1976). The differentiation of material lifestyles: 1925–1966. *Social Indicators Research, 3*, 397–421.

Fournier, S., Antes, D., & Beaumier, G. (1992). Nine consumption life-styles. *Advances in Consumer Research, 19*, 329–337.

Gans, H. (1968). Urbanism and suburbanism as ways of life. In R. Pahl (Ed.), *Readings in urban sociology* (pp. 95–114). London: Pergamon.

Ganzeboom, H. (1988). *Leefstijlen in Nederland. Een verkennende studie*. Sociaal en Cultureel Planbureau. Cahier no. 60. Alphen aan den Rijn: Samson Uitgeverij.

Ganzeboom, H. (1990). Leefstijlen: een theoretische en methodologische beschouwing met een empirisch voorbeeld. In *Jaarboek van de Nederlandse Vereniging van Marktonderzoekers* (pp. 89–90). Haarlem: De Vrieseborch.

Gibler, K. M., & Nelson, S. L. (2003). Consumer behaviour applications to real estate education. *Journal of Real Estate Practice and Education, 6*(1), 63–83.

Grünfeld, J. (2003). *Gestrand op Kanaleneiland. Woonsatisfactie en verhuisgeneigdheid van Turkse, Marokkaanse en Nederlandse bewoners in een herstructureringswijk te Utrecht*. Thesis, University of Amsterdam, Amsterdam .

Gunter, B., & Furnham, A. (1992). *Consumer profiles. An introduction to psychographics*. London: Routledge.

Haest, G. (1989). *De oude garde, het andere slag en de buitenlanders. De geschiedenis van een saneringswijk*. Assen: Van Gorcum.

Hagen, G. J. (2006). Leefstijlen. De klant in de mand... *Boss Magazine* 22–27.

Hamilton-Smith, E. (2000). *Lifestyles: Patterns and trends* (Research discussion paper no 13). Melbourne: University of Melbourne, Centre for Applied Research on the Future.

Heijs, W., Carton, M., Smeets, J., & van Gemert, A. (2005). *Labyrinth van leefstijlen*. Eindhoven: Cahier Architectuur Stedebouw Eindhoven.

Heijs, W., Carton, M., Smeets, J., & van Gemert, A. (2009). The labyrinth of life-styles. *Journal of Housing and the Built Environment, 24*(3), 347–356.

Hessing, E., & Reuling, A. M. H. (2003). Waarden in Nederland. Segmentatie van doelgroepen. In A. E. Bronner, P. Dekker, J. C. Hoekstra, E. de Leeuw, T. Poiesz, K. de Ruyter, & A. Smidts (Eds.), *Ontwikkelingen in het Marktonderzoek: Jaarboek MarktOnderzoeksAssociatie* (pp. 163–175). Haarlem: Spaar en Hout.

Hooimeijer, P., & Schutjens, V. (1991). Changing lifestyles and housing consumption: A longitudinal approach. *Journal of Housing and the Built Environment, 6*(2), 143–158.

Hustad, P., & Pessemier, E. (1974). The development and application of psychographic life-style and associated activity and attitude measures. In W. Wells (Ed.), *Life-style and psychographics* (pp. 32–70). Chicago: American Marketing Association.

Inbo. (2009 September). *Nieuwsbrief Adviseurs*, 34.

Jansen, S. J. T., Coolen, H. C. C. H. (2010). *What is the worth of lifestyle variables in the prediction of preferences for private and public green spaces?* Working paper, presented at the 22nd conference of the European Network for Housing Research (ENHR), Istanbul.

Kahle, L. R. (1983). *Social values and social change: Adaptation to life in America*. New York: Praeger.

Kersloot, J., & Kauko, T. (2004). Measurement of housing preferences–A comparison of research activity in the Netherlands and Finland. *Nordic Journal of Surveying and Real Estate Research, 1*, 144–163.

Kleinhans, R., & Elsinga, M. (2010). 'Buy your home and feel in control' does homeownership achieve the empowerment of former tenants of social housing? *International Journal of Housing Policy, 10*(1), 41–61.

Kolpron Consultants and MarketResponse. (2000). *Rapport Woonbeleving 2000*. Amersfoort: The SmartAgent Company.

Kruskal, J. B. (1964). Multidimensional-scaling by optimizing goodness of fit to a Nonmetric Hypothesis. *Psychometrika, 29*(1), 1–27.

Merton, R. K. (1957). *Social theory and social structure*. Glencoe: Free Press.

Michelson, W., & Reed, P. (1974). Lifestyle in environmental research. In C. Beattie & S. Crysdale (Eds.), *Sociology Canada: Readings* (pp. 406–419). Toronto: Butterworths.

Mitchell, A. (1983). *The nine American life styles*. New York: Warner.

Nijhuis, M., & Schoemaker, R. (2002). Normen, waarden en woonbelevingsgroepen. Consumentenonderzoek van Motivaction. *Stedebouw en Ruimtelijke Ordening, 6*, 43–45.

Overboom, L. (2003). *Tussen klant en stad.* Cahier reeks Duurzame stedelijke vernieuwing (2(11)). KEI Kenniscentrum Stedelijke Vernieuwing en NIDO Programma Duurzame Stedelijke Vernieuwing.

Pinkster, F., & van Kempen, R. (2002). *Leefstijlen en Woonmilieuvoorkeuren*. Utrecht: University of Utrecht, Urban and Regional Research Centre.

Reijndorp, A., Kompier, V., & de Haas, L. (Eds.). (1997). *Wonen in de 21e eeuw. Leefstijlen*. Rotterdam: NAi.

Reijndorp, A., Kompier, V., Metaal, S., Nio, I., & Truijens, B. (1998). *Buitenwijk. Stedelijkheid op afstand*. Rotterdam: NAi Uitgevers.

Rokeach, M. (1973). *The nature of human values*. New York: Free Press.

Rossi, P. H. (1955). *Why families move: A study in the social psychology of urban residential mobility*. Glencoe: Free Press.

Schwartz, S. H. (1992). Universals in the content and structure of values: Theoretical advances and empirical tests in 20 countries. In M. P. Zanna (Ed.), *Advances in experimental social psychology* (Vol. 25, pp. 1–65). San Diego: Academic.

Schwartz, S. H., & Boehnke, K. (2004). Evaluating the structure of human values with confirmatory factor analysis. *Journal of Research in Personality, 38*, 230–255.

Shevky, E., & Bell, W. (1955). *Social area analysis*. Stanford: Stanford University Press.

Sobel, M. E. (1981). *Lifestyle and social structure: Concepts, definitions, analyses*. New York: Academic.

SRI International. (1989). *VALS 2*. Menio Park: SRI International.
Tallman, I., & Morgner, R. (1970). Life-style differences among urban and suburban blue-collar families. *Social Forces, 48*(3), 334–348.
Urhahn Urban Design. (1999). *Suburbaan wonen in de stad. Een onderzoek naar de kwaliteit en duurzaamheid van woonmilieus in Apeldoorn.*
van der Lelij, B. (2002). Mentality: Vernieuwde normen- en-waardensegmentatie. *eYe. Zicht Op Trends, 4*(6/7), 14–15.
van der Schaaf, R. (2002). Leefstijlen en woonmilieus in de wijkvernieuwing. De Groninger methode. *Stedebouw & Ruimtelijke Ordening, 6*, 40–42.
van der Wouden, R., & Kullberg, J. (2002). Stijloefeningen. Leefstijlen in onderzoek en praktijk. *Stedebouw & Ruimtelijke Ordening, 6*, 13–20.
van Diepen, A. M. L., Arnoldus, M. (2003). *De woonvraag in de vraaggestuurde markt. Bouwstenen uit het woonmilieuanalyse en leefstijlenonderzoek* (rapport 23). Utrecht: DGW/Nethur partnership.
van Diepen, A., & Musterd, S. (2001). *Stedelijke leefstijlen en woonmilieus in Amsterdam.* Amsterdam: Amsterdam study centre for the Metropolitan Environment.
van Diepen, A. M. L., & Musterd, S. (2009). Lifestyles and the city. *Journal of Housing and the Built Environment, 24*, 331–345.
van Engelsdorp-Gastelaars, R. (1980). *Niet elke stadsbewoner is een stedeling*. Amsterdam: Universiteit van Amsterdam, Sociaal-Geografisch Instituut.
van Kralingen, R. (2003). *Super brands. Merken en markten van morgen*. Deventer: Samson.
Veal, A. J. (2000). *Leisure and lifestyle: A review and annotated bibliography* (Online Bibliography 8). Sydney: University of Technology, School of Leisure, Sport & Tourism.
Vyncke, P. (2002). Lifestyle segmentation: From attitudes, interests and opinions, to values, aesthetic styles, life visions and media preferences. *European Journal of Communication, 17*, 445–463.
Weber, M. (1922). Wirtschaft und Gesellschaft. Tübingen: Mohr.
Wedel, M., & Kamakura, W. A. (2000). *Market segmentation: Conceptual and methodological foundations* (2nd ed.). Norwell: Kluwer Academic Publishers.
Wells, W. (Ed.). (1974). *Life-style and psychographics*. Chicago: American Marketing Association.
Wirth, L. (1938). Urbanism as a way of life. *The American Journal of Sociology, 44*(1), 1–24.
Zablocki, B. D., & Kanter, R. M. (1976). The differentiation of life-styles. *Annual Review of Sociology, 2*, 269–298.

Internet References

De Bock and Dekker. (2009). http://www.d-sciencelab.be/newpic/glossary/get_file.cfm?file=segm_De_Bock.htm
Inbo and I&O Research. (2009). http://www.enschede.nl/gemeente/actueel/stedelijkcie/00080/11579_woonvisie_bijl.3_woonmilieus.doc
National Readership Survey. (2009). http://www.businessballs.com/demographicsclassifications.htm
NFO Trendbox. (2009). http://www.cos.nl/onor/archief/leefst/combur.pdf
Stuurgroep Experimenten Volkshuisvesting. (2009). http://www.platformgras.nl/showfile.php?c=download&id=21
Value Group Ltd. (2009). http://www.businessballs.com/demographicsclassifications.htm
Van der Flier Consultancy. (2009). http://www.reurba.org/reports4/lifestyles-annex.html
Young and Rubicam. (2009). http://www.4cs.yr.com/public

Chapter 9
Neo-classical Economic Analysis

Marnix J. Koopman

9.1 Introduction

In this chapter the neo-classical view of the housing market and two revealed preference models that can be derived from it are explained. Neo-classicists make strict assumptions about the behavior of households and producers that are probably not met in real life. Their equilibrium framework is however of great practical relevance. This means that researchers and professionals in the housing market must know the raw basics behind the neo-classical theory and techniques.

Central to the neo-classical theory is the axiom of revealed preferences. It states that buyers and sellers are able to rank and value the bids and offers for goods on the market. The subjective value that households attach to a good gives rise to their bids. The exchange of goods (i.e. dwellings) only takes place among buyers who cannot find another seller asking for less, and sellers who cannot find another buyer who will bid higher. The optimal choices of sellers and buyers on the housing market can thus reveal their preferences for housing quality.

This chapter is organized as follows. After examining the theory behind neo-classical housing economics in Sect. 9.2, two commonly used revealed preference models will be examined in Sect. 9.3. Both the likelihood to move of tenants in Sect. 9.4 and the hedonic house price equation in Sect. 9.5 are explained to highlight their practical value. In Sect. 9.6 some limitations of the neoclassical framework are discussed, and how they are dealt with in the literature. Suggestions for further reading are also given that go beyond the simple approach taken in this chapter.

M.J. Koopman (✉)
Netherlands Environmental Assessment Agency (PBL), OTB Research Institute,
Delft University of Technology, Bilthoven, The Hague, The Netherlands
e-mail: m.j.koopman@tudelft.nl; marnix.koopman@pbl.nl

S.J.T. Jansen et al. (eds.), *The Measurement and Analysis of Housing Preference and Choice*, DOI 10.1007/978-90-481-8894-9_9,

9.2 The Neoclassical Housing Market

In the neoclassical view of the housing market, price and rent adjust over time and in space to create long-run equilibrium between the supply and demand for dwellings. The decision to leave or improve one's home, to choose a new home in order to relocate, to move from renting to owner-occupation and vice versa make up the household's choice set. On the supply-side the landlord buys, lets or sells property, the constructor demolishes, renovates and constructs dwellings and the developer acts as the middle man. Aggregation of net construction (construction minus demolition), exit and relocation rates create the housing supply and demand. The equality of demand and supply in turn yields the equilibrium house price and rent.

9.2.1 Market Equilibrium

On the supply-side, arbitrage – the unlimited ability to buy a good and sell it on a different market – equalizes landlords' profits from selling or letting property. If activities on the housing market are more profitable than holding other assets, landlords – and developers if dwellings are sold directly to households – commission firms to build dwellings. The change in the housing stock is the surplus of construction over demolition. The stock is fixed in the short run due to the time lag in construction. In the long run a higher price and rent lead to more construction, so that long-run supply is an upward-sloping function of the rent and price.

Housing demand is defined by the number of households in the area, which changes as a result of net relocation (inward migration and net household formation minus outward migration and deaths) to the area. A higher house price or rent reduces a households' desire to stay in or move to an area, so that demand is a downward-sloping function of the rent and price. The equality of supply and demand determines the equilibrium rent and price at the equilibrium housing stock.

The equilibrium can be extended to each attribute of the dwelling. An attribute's contribution to the house price is the maximum amount that a household is willing to pay for that particular item of housing quality. Furthermore, households and producers face the same exchange value for the attribute, regardless of their individual preferences. The equilibrium price P of dwelling i thus depends on the quantity of each of the N attributes X_n of the dwelling and neighborhood alone. The function that explains the level of house prices (and market rents) can be written as:

$$P^{i_i} = f(X_1^i, X_2^i, ..., X_N^i) \tag{9.1}$$

9.2.2 Market Dynamics

The neoclassical framework is based on equilibrium analysis, while its dynamics (i.e. the housing choices) result from adjustments towards the equilibrium. Having children induces the household to move to a larger dwelling that is optimal from its new standpoint. A large number of births increase the demand for space at every price level. Excess demand gives rise to a disequilibrium in which the price rises. If the rise in demand is temporary, no construction takes place and both demand and price return to their former values. If construction takes place, the size of the stock increases over the course of time. This leads to a new equilibrium, where the price is higher than in the old equilibrium, but lower than in the temporary disequilibrium (see Fig. 9.1).

Households act on their own accord and they take the equilibrium price and stock as the starting point of their housing choice, so that the market as a whole ends up in equilibrium. With equilibrium prices and quantities in mind, households do not concern themselves with past decisions, decisions that are taken in the future or the alternatives that are discarded when making the choice. The exit choice is the household's optimal response to its current housing situation. The probability of a move by household *i* depends on the *M* characteristics of the household HH_m that define its taste for housing quality, the house price *P* (or rent) and the *L* attributes X_l of the current home and neighborhood.

$$\Pr^i(exit) = g(X_1^i, X_2^i, \ldots X_L^i; HH_1^i, HH_2^i, \ldots HH_M^i; P^i) \tag{9.2}$$

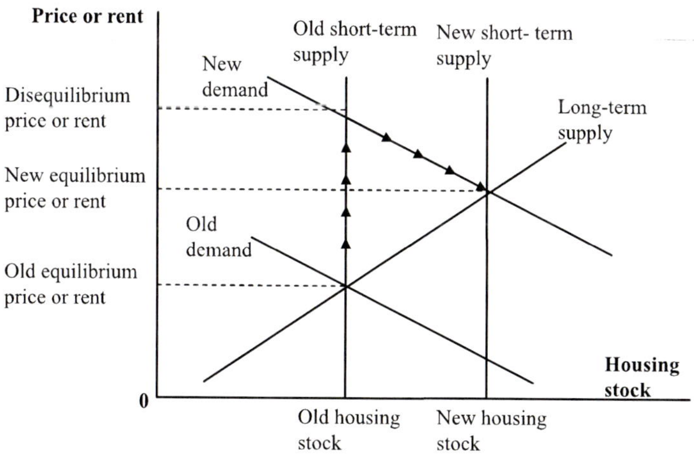

Fig. 9.1 Adjustment towards equilibrium on the housing market

9.2.3 *Neoclassical Assumptions*

Neo-classicists assume rational behavior, full information and households and producers that act on their own accord. Although rationality is not always guaranteed, if the prices are the same few households will prefer an inferior dwelling. Complete information and atomistic actions are far less realistic assumptions. Households can only rank dwellings that they are aware of and even then they seldom know the exact quality of an offer. Cooperation is a fact of life on the supply-side, just think about spatial planning or urban restructuring. Nevertheless the relations that are derived from the neoclassical framework are extremely useful in practical applications, especially if the market contains elements of free exchange.

9.2.4 *Why and When to Apply the Neoclassical Approach*

The closer a market relation resembles the free exchange of goods where there is complete information, the more faith can be put in the revealed preference model that explains it. Housing demand is easier to explain than housing supply, because land-use restrictions impede suppliers' ability to freely build where they want to. Demand in the owner-occupied market is easier to model than the rental market, because rent controls distort the free exchange between landlords and tenants. It is harder to model the choice of a new home than the decision to leave the old home, because house-seekers know less about their new home than residents know about their current home.

As a minimal requirement for the use of revealed preference models households must be able to rank the dwelling that they are selling, buying or renting, and they must have the option to forego an offered dwelling and/or select another one. The price relationship in Sect. 9.1 withstands this heuristic test. On the owner-occupied market buyers and sellers know the price and structural characteristics of the dwelling and they can always forego an offer or decline a bid. The distribution system of social rented dwellings might not offer the same opportunities as the owner-occupied market, but a tenant has the option to select a new rented dwelling or to stay in his current dwelling and wait until a better offer comes along. Hence a revealed preference model for the exit choice of social tenants does have real meaning.

9.3 Methodology

9.3.1 *Modeling Market Dynamics and Equilibrium*

In two fields of research the applications of the neoclassical framework have been extremely successful, despite the fact that free housing choice and full information are far from certain. I will first use a discrete choice model to estimate the likelihood that a social tenant will leave his dwelling. This highlights the usefulness of revealed preference models even in a regulated market. I will then consider the most popular revealed preference model for the housing market: a hedonic house price equation for the same city. It is assumed that the reader has had some experience with regression analysis. For readers with more statistical experience, it is an opportunity to get (re-)acquainted with the potential pitfalls of revealed preference modeling.

9.3.2 Data

The data that is used in the two models was collected from the housing stock of the Woonbron housing association in Rotterdam. The records of the housing association provide detailed data on the structural characteristics and rent of dwellings, some limited data on tenants at the beginning of 2002, and the assessment value of the dwellings in 1999. House values are discussed and used in Sect. 9.4. The occurrence of households leaving the rental sector over a 3 year period (2002, 2003 and 2004) is also known. Although the reason for leaving is unknown, it is assumed that a sizeable number of these exits were voluntary. The data from the housing association are appended by some neighborhood statistics from the municipal statistics office in Rotterdam.

Which structural characteristic (house type, house size, building period, number of rooms etc.), household characteristics (age of the head of the household, duration of stay, income etc.) and neighborhood attributes should enter the exit choice model, depends on data availability and the question the researcher poses. The inclusion of more structural characteristics than listed here (floor level, garage, garden, number of bathrooms, type of heating etc.) can only improve a housing market model, in that each item of housing quality is catered for. Household characteristics can be used to reveal the preferences of different categories of household for items of housing quality, or they can answer a specific question. The researcher might for instance want to know what effect the age, getting married or commuting time has on residential mobility.

A researcher might further want to know what impact the aspects of the surroundings have on the exit choice and housing quality in general. The statistics on neighborhood attributes fall into the broad categories of accessibility (distance or travel-time to city centre, railway station or highway), physical (dominant house type in area, presence of shops and parks), demographical (household type, age, size), socio-economical (owner-occupation and unemployment in the area, income, education, type of job), cultural (ethnicity, land of origin), social-environmental (crime and victimization rate), institutional-environmental (school quality and local tax rates) and ecological environmental (noise and pollution levels in the area, flood zones) attributes.

Two types of neighborhood statistics can be identified. First some statistics are centered at the amenity. The distance or travel-time from the dwelling to the amenity, or the value of the spatial rate (e.g. noise, pollution or social disturbance rate) at the location of the dwelling can then be calculated. The second type is the mean of observations inside some predefined range. If micro-data and coordinates are available, this average can be centered at the dwelling, with higher weights given to observations that are closer to the dwelling than to observations that are further away. Most statistical agencies calculate statistics as the mean of observations inside some predefined spatial unit (e.g. postal code area, census tract). In the latter case the degrees of freedom are only as high as the number of spatial units in the dataset rather than the number of observations. I restrict the use of the neighborhood attributes to just three: distance to the city-centre, and two neighborhood means that are commonly used in housing market models: the percentage of owner-occupied dwellings and average household income.

9.3.3 Techniques

The revealed preference models of the housing market are estimated with the aid of regression techniques. The starting point is a presumed causal relationship between a dependent variable and one or more explanatory variables. The dependent variable is then regressed on the variables based on a large number of observations. As an inductive research method the regression results are supposed to apply to the population as a whole, even though the estimation only takes place for a sample of households and/or dwellings. The estimation outcome reveals whether a variable has an impact on the dependent variable and what the exact size (or monetary value) of that impact is, when the values of the other explanatory variables are left unchanged.

The dependent variable in this exit choice model is the likelihood to move. This is the probability that a household exits within a given period of time, based upon actual exits and stays in a similar time frame. In this case I use a 3 year period, but any period from 1 to 5 years will do. It is not advisable to extend this period any longer as many of the data change value over time, but more importantly the relation can undergo a structural change. The likelihood to move is based on a discrete choice, since exits and stays are mutually exclusive. Common convention dictates that an exit receives the value 1 and a stay the value 0. The easiest way to model a discrete choice is a logit-model that is available in most statistical software packages. There exists a wealth of regression techniques for the house price equation in Sect. 9.5. For the sake of brevity, only the simplest one will be considered here: Ordinary Least Squares estimation of the logarithm of prices. More advanced methods for both models are discussed in Sect. 9.6.

9.4 Exit Choice

9.4.1 Estimation of the Likelihood to Move: Concept and Data

Households must at some stage during their housing career decide if they want to stay or move. The exit choice model predicts what the most likely form of action is going to be based on the individual characteristics of the household and the dwelling in which they live. One hundred observations are seen as the absolute minimum for the estimation of discrete choice models. The larger the sample the more likely it is that the sampled households act in the same way as the population as a whole. With plenty of observations, you can also leave some of them for later to test the predictive powers of the model.

As mentioned before, the exit choice is a function of the price or rent of the dwelling, the attributes of the dwelling and neighborhood and the household's characteristics. The coefficients of the structural characteristics indicate whether the average household's exit behavior is affected by living in a certain type of dwelling. Household characteristics control for the individual taste for housing quality. Together these explanatory variables can reveal what the preferences of different categories of households are for selected items of housing quality. Adding neighborhood attributes or neighborhood dummies to the model controls for area-based differences in the exit rates and spatial interactions among the explanatory variables.

Listed below in Table 9.1 are the data that are used in the estimation of the exit choice model, their sample mean, standard deviation and average rate of exit for each category of the explanatory variable. Some explanatory variables come in a continuous form. Very high or low values in these variables can have a dramatic effect on the estimation. Observations that fall outside a certain range around the mean of a continuous variable should be dropped from the sample.[1] This applies to very expensive or cheap dwellings (in this case a rent lower than 150 or higher than 750 Euros) and households with a very low duration of stay (less than 1 year). Households with a low duration of stay should not be sampled anyway, because they seldom move and if they do, it is the result of a suboptimal choice (a repeat move) or an involuntary one (divorce or eviction), both of which are best not investigated with a revealed preference model.

Applying the natural logarithm is a common transformation for continuous variables. A small change in a logarithmic variable equals the percentage increase in that variable. As such the coefficient measures the percentage impact that the variable has on the dependent variable. As an alternative, continuous variables can be written as a combination of linear and quadratic terms. The best type of specification is more a matter of test by trial and error than of any fixed rules. In the exit choice model below rent, duration of stay and neighborhood income are all specified as logarithms, while the age of the head of the household is written as age and age squared.

Missing values on continuous variables can be replaced by the sample mean of that variable or by the predicted values from a regression of this continuous variable on the other explanatory variables. The larger your dataset is the less necessary it becomes to retain extreme values, observations with missing values and other polluted data.

---

[1] One way of doing this is by sampling the range within the mean of the continuous variable minus 1.96 times its standard deviation and the mean plus 1.96 times its standard deviation. For a normally distributed variable this range contains about 95% of the observations.

Table 9.1 Dataset likelihood to move

			Mean	St. dev	% exits
Structural characteristics	Building period	Built before 1945	0.215	0.411	19.3
		Built 1945–1975	0.357	0.479	19.9
		Built 1975–1990	0.376	0.484	22
		Built after 1990	0.043	0.202	11.8
		Unknown	0.009	0.095	22.2
	House type	Multi-family downstairs	0.139	0.346	12.5
		Multi-family upstairs with lift	0.228	0.42	20.8
		Multi-family upstairs no lift	0.409	0.492	19.4
		Single-family	0.222	0.416	24.5
		Unknown	0.001	0.035	38.5
	Number of rooms	One	0.015	0.123	51.4
		Two	0.143	0.35	29.5
		Three	0.404	0.491	21.4
		Four	0.286	0.452	15.1
		Five or more	0.151	0.358	13.2
	Size	Less than 50 m^2	0.188	0.391	29.6
		Between 50 and 60 m^2	0.259	0.438	21.1
		Between 60 and 75 m^2	0.384	0.486	18.4
		Between 75 and 90 m^2	0.149	0.356	14.6
		More than 90 m^2	0.02	0.14	11.0
Housing costs	Rent	Rent in Euros	372.06	90.56	–
Household characteristics	Income category	Low	0.986	0.119	19.7
		High	0.014	0.119	52.8
	Age of head	Age in years	56.415	16.813	–
	Duration of stay	Stay in months	127.67	102.72	–
Neighborhood attributes	Accessibility	Distance to city-centre in km.	4.89	1.793	–
	Socio-economical	Average household income in neighborhood	17,207	1,736.75	–
		% of owner-occupation in neighborhood	19.983	12.46	–

Number of observations = 8,845

Categorical variables, which have been transposed to dummy variables, have the value *1* if the observation falls within the specific category and *0* otherwise. Some structural characteristics are categorical by definition: house type and the presence of a lift for instance. In the present example, the categorical nature of the building period, number of rooms and house size, was more a matter of data availability than a deliberate choice.

Ironically, it can be better to have more rather than a few missing values on a categorical variable. With a lot of missing values a new category can be created, which contains all the observations of the explanatory variable with missing values. With only a few missing values the observations must be dropped from the estimation.

Estimation of models with categorical explanatory variables must take place with one category of each explanatory variable as part of the frame of reference. Failing to do so makes estimation impossible, because some of the explanatory variables are self-explanatory (i.e. each referential category equals one minus the values of the other categories). The common value of all the referential categories is contained in the constant term of the equation. Four common problems with categorical variables are that:

1. Nearly all the observations fall into a single category of an explanatory variable. The near empty categories can be dropped and estimation must take place for the observations in the prevalent category only, and without the explanatory variable.
2. Few observations fall into a category. Smaller categories might be joined with an adjacent category of the same explanatory variable.
3. The category is a (nearly) perfect predictor. Even though this is an interesting outcome from an analytical viewpoint the regression has no meaning. The observations in that specific category must be dropped; it is already known what the impact of that category is going to be.
4. One category overlaps with the category of another explanatory variable. This creates multicollinearity: one explanatory variable (almost) explains another one. All observations can be retained but either one of the categories will not enter the model. The coefficient on that category estimates the inextricable effect of both categories.

9.4.2 *Modeling the Likelihood to Move Among Social Renters in Rotterdam*

The above data was used to model and estimate formula 2 for social renters in Rotterdam in the years 2002, 2003 and 2004. Table 9.2 lists the coefficients, z-values and 95%-confidence intervals for the coefficients in a robust estimation of a logit-model for 1,770 exits (20.0%) among 8,845 social renters. The frame of reference, which is contained in the constant, is a fictional 1-room downstairs apartment, built before 1945, smaller than 50 m² and inhabited by a low income tenant. All coefficient values are based on the 'average household' with an age of 56 years and duration of stay of 128 months. The estimation results are interpreted in three ways: the odds-ratios, marginal effects and Willingness-To-Pay.

The choice of the logit model is suggested by a handy feature of this model. In the estimated equation the coefficient values on categorical variables can be turned into odds-ratios by taking their exponential value:

$$\frac{\Pr(Y^{i_t} = 1; X_{rj})}{\Pr(Y^{i_t} = 1; X_{0j})} = \exp(\beta_{rj}) \qquad (9.3)$$

Table 9.2 Estimation results for the likelihood to move among social renters 2002–2004

			Coefficient	Z-value	Lower bound	Upper bound
	Constant		−7.91	−1.24	−20.39	4.58
Structural characteristics	Age	Built before 1945 (reference)				
		Built between 1945 and 1975	−0.24	−2.15	−0.46	−0.02
		Built between 1975 and 1990	−0.78	−7.15	−0.99	−0.57
		Built after 1990	−1.60	−8.48	−1.97	−1.23
		Building period unknown	0.38	0.95	−0.40	1.15
	House type	Multi-family downstairs (reference)				
		Multi-family upstairs with lift	−0.54	−4.81	−0.75	−0.32
		Multi-family upstairs without lift	0.42	4.52	0.24	0.60
		Single-family	−1.27	−9.25	−1.54	−1.00
		House type unknown				
	Number of rooms	One (reference)				
		Two	−1.80	−8.28	−2.23	−1.37
		Three	−2.26	−9.62	−2.72	−1.80
		Four	−2.83	−10.86	−3.34	−2.32
		Five and more	−3.03	−9.74	−3.63	−2.42
	Size	Less than 50 m² (reference)				
		Between 50 and 60 m²	−1.13	−9.08	−1.37	−0.88
		Between 60 and 75 m²	−1.49	−9.23	−1.81	−1.17
		Between 75 and 90 m²	−1.84	−8.28	−2.27	−1.4
		More than 90 m²	−2.59	−7.05	−3.31	−1.87

Rent		Logarithm of rent	5.51	17.55	4.90	6.13
Household characteristics	Income category	Low (reference)				
		High	0.64	3.07	0.23	1.04
	Age of head	Age	–0.16	–13.34	–0.18	–0.14
		Age squared	0.00121	12.51	0.00102	0.00140
	Duration of stay	Logarithm of duration of stay	0.12	2.74	0.03	0.20
Neighborhood attributes	Distance to city-centre	Distance in km	0.0639	2.35	0.0111	0.12
	Neighborhood income	Logarithm of average income	–1.85	–2.87	–3.12	–0.59
	Owner-occupation	% of owner-occupied stock	0.00117	0.22	0.000914	0.01150
Pseudo R^2	0.132	LR-statistic	1167.9			
N	8,845	Probability(LR.)	0.000			

for category r of structural characteristic X_j as opposed to the frame of reference. You can also calculate the odds that a specific category r of household characteristic j and category s of structural characteristic h moves as opposed to the referential categories through multiplication (i.e. $exp(\beta_{rj})*exp(\gamma_{sh})$). Or you can calculate the odds that a category r of household characteristic j moves as opposed to another category s of household characteristic j through division (i.e. $exp(\beta_{rj})/exp(\beta_{sj})$). For continuous variables the coefficient must also be multiplied by the change in that variable. If this change equals 1, formula 9.4 applies (see below).

So from Table 9.2 the odds-ratio of living in a single-family dwelling as opposed to a downstairs apartment is the exponential value of the coefficient on that specific category: exp(−1.27) = 0.28, with an upper bound of 0.37 (=exp(−1.00)) and lower bound of 0.22 (=exp(−1.54)). The odds-ratio of 0.28 can be interpreted as residents in a single-family dwelling being 0.28 times as likely to move as residents living in a downstairs apartment. In other words, residents living in a single-family dwelling are 1/0.28 = 3.6 times less likely to move than respondents living in a downstairs apartment.

The odds that an average high-income household in a single-family dwelling moves compared with the frame of reference follows from multiplication. The exponential value of the coefficient for the high income category is multiplied by the odds that were obtained above. This odds-ratio equals exp(0.64)* 0.28=1.89* 0.28 = 0.53. There is a 47% lower chance that the high-income household in a single-family dwelling moves, compared with the low-income household in a downstairs apartment.

Likewise division can be used to see how a household fares in an upstairs apartment with a lift as opposed to when it lives in a single family dwelling. The odds-ratio for the change of house type is the exponential value of the coefficient for an upstairs apartment with a lift, divided by the exponential value of the coefficient for the single-family dwelling category or 2.1 (=0.58/0.28 = exp(–0.54)/exp(−1.27)).

While odds-ratios are often used in sociological and psychological studies, economists prefer to work with the marginal effect: the change in the likelihood to move due to a small change in the explanatory variable. The marginal effect is given as:

$$\frac{\delta \Pr(Y^i = 1; X_{kq})}{\delta X_{kq}} = \frac{\beta_{kq} * \exp(-\alpha - \sum \beta_{kl} X^i_{kl} - \beta_p \ln(P^i) - \sum \chi_{km} HH^i_{km})}{(1 + \exp(-\alpha - \sum \beta_{kl} X^i_{kl} - \beta_p \ln(P^i) - \sum \chi_{km} HH^i_{km}))^2} \tag{9.4}$$

The left-hand side of the equation is the derivative of the likelihood to move with respect to variable X_q, or the effect that a small change in that explanatory variable has on the likelihood to move. In the case of a categorical variable this is the change in the probability due to the change of category from r to k: $Pr(Y^i = 1;X_{kq})-Pr(Y^i = 1;X_{rq})$. The right-hand side can be rewritten as $\beta_{kq}* (1-Pr(Y^i = 1;X_q))*Pr(Y^i = 1;X_q)$, or the coefficient on X_q times the likelihood to stay times the likelihood to move evaluated at some chosen value of X_q.

The marginal effect can be evaluated at any given value for the explanatory variables. Statistical software packages give the option to save the predicted probabilities of discrete choice models. The predicted likelihood that the average low-income family living in a downstairs apartment will move is 0.13 (nearly all software packages have the option to display and store the predicted values from the regression). The marginal effect on the likelihood to move for a change from the downstairs apartment and low-income category to a single-family dwelling and high-income category is then 0.075 or (0.64–1.27)*(1–0.13) * 0.13. The likelihood to move thus drops from 0.13 to 0.12 when you move from the frame of reference to the high income and single family dwelling categories.

The last application of the exit choice model is the Willingness-To-Pay. According to theory, combinations of the explanatory variables that keep the likelihood to move constant, keep the household on its optimal consumption path. The Willingness-To-Pay is the amount of rent needed to keep the likelihood to move constant when structural characteristic X_s changes.

$$\beta_p \delta \ln P + \beta_{ks} \delta X_{ks} = 0 \Rightarrow WTP = \frac{\delta P}{\delta X_{ks}} = -\frac{\beta_{ks}}{\beta_p} * \overline{P} \tag{9.5}$$

The Willingness-To-Pay equals the negative value of the coefficient β_{ks} on the structural characteristic, divided by the coefficient on rent β_p, multiplied by the average rent $\overline{P}$. Multiplication by the average rent is needed because of the logarithmic transformation of the rent (i.e. $\delta ln(P)/\delta P = 1/P$).

The sample mean of the rent ($\overline{P}$) is € 372.06 (see Table 9.1). According to formula 9.5 the amount of money that the average household is willing to pay for a change from a downstairs apartment to a single-family equals the negative value of the coefficient on single-family dwelling divided by the coefficient on the rent times the average rent, or: € 85.57 (=–(−1.27)/5.52) * € 372.06).

9.5 House Prices

9.5.1 *Estimation of the Hedonic House Price Equation: Concept and Data*

The hedonic pricing method, though certainly not the only valuation method in use, has become the main tool in the assessment and prediction of house prices (and market rents). This method states that price differences can be explained by the structural characteristics of dwellings and their neighborhood attributes alone. In theory the hedonic price function reveals the maximum amount that buyers wish to pay for housing attributes.

If a hedonic price equation is available for a particular housing market, filling in the values of known housing attributes reveals what the price of the dwelling is likely to be, even before it has been built. It thus tells producers what the expected revenues are from constructing a new dwelling or improving an existing dwelling. Real estate agents, municipal tax and land registry offices also make frequent use of hedonic models in mass appraisals.

Table 9.3 Data hedonic models

		Mean	Standard deviation
House price	1999	65,932	15,826
Age	Built before 1945 (reference)	0.23	0.42
	Built between 1945 and 1975	0.37	0.48
	Built between 1975 and 1990	0.33	0.47
	Built after 1990	0.05	0.23
	Building period unknown	0.01	0.11
House type	Multi-family downstairs (reference)	0.16	0.37
	Multi-family upstairs with lift	0.34	0.47
	Multi-family upstairs without lift	0.42	0.49
	Single-family dwelling	0.08	0.27
	House type unknown	0.001	0.04
Number of rooms	One (reference)	0.02	0.15
	Two	0.22	0.41
	Three	0.43	0.49
	Four	0.28	0.45
	Five and more	0.05	0.22
House size	Less than 50 m² (reference)	0.27	0.44
	Between 50 and 60 m²	0.29	0.45
	Between 60 and 75 m²	0.37	0.48
	Between 75 and 90 m²	0.07	0.25
	More than 90 m²	0.01	0.11
Number of observations		13,564	

Here the sample consists of a set of 13,564 dwellings in 1999, with a value less than 100,000 Euros. The hedonic house price equations uses the same structural characteristics as those that were used in the exit choice model: house type, building period, number of rooms and size of the dwelling. Not listed in Table 9.3 are the area-based dummies that receive the value 1 if the dwelling lies in a specific neighborhood of the city and 0 elsewhere. There are a total of 25 of these dummy variables for the 26 administrative neighborhoods in the dataset, with the observations in one neighborhood as part of the frame of reference. Neighborhood dummies control for area-based differences in the house price and spatial interactions among the explanatory variables (see Sect. 9.6). This ensures that the coefficients on the structural characteristics measure what they are supposed to do: the exchange value of the dwelling's attributes on the owner-occupied market.

9.5.2 *Modeling the Hedonic House Price Equation for Rotterdam*

Based upon the technical notions above a hedonic house price model for the (logarithm of the) assessment values of 13,564 dwellings in 1999 in Rotterdam can be estimated. The outcome of these regressions by ordinary least squares is given in Table 9.4. First a few price elasticities are calculated and then a robustness check is

Table 9.4 Estimation results for the hedonic house price in 1999

		Coefficient	T-value
Constant		10.43	986.92
Age	Built before 1945 (reference)		
	Built between 1945 and 1975	0.02	4.06
	Built between 1975 and 1990	0.16	30.40
	Built after 1990	0.25	43.68
	Building period unknown	–0.06	–5.39
House type	Multi-family downstairs (reference)		
	Multi-family upstairs with lift	0.07	18.04
	Multi-family upstairs without lift	–0.07	–20.27
	Single family dwelling	0.30	49.87
	House type unknown	–0.01	–0.46
Number of rooms	One (reference)		
	Two	0.35	43.36
	Three	0.52	57.83
	Four	0.58	58.94
	Five and more	0.62	50.86
Size	Less than 50 m² (reference)		
	Between 50 and 60 m²	0.08	17.43
	Between 60 and 75 m²	0.17	28.73
	Between 75 and 90 m²	0.23	27.33
	More than 90 m²	0.30	23.15
Neighborhood dummies included		Yes	
		$R^2=0.75$	
		F-statistic = 1010.5 (p = 0.000)	
		N = 13.564	

carried out for the model that is applicable to all revealed preference models: a test of its predictive powers.

There are two operations that researchers usually carry out with the hedonic equation. The first is the calculation of the price elasticity of a variable:

$$\frac{\delta P}{\delta X_{k,n}} \frac{\overline{X}_{k,n}}{\overline{P}} = \beta_{k,n} * \overline{X}_{k,n} \tag{9.6}$$

The left-hand side of the equation can be rewritten as $(\delta P/P)/(\delta X_n/X_n)$ or the relative increase (or decrease) in the house price for a relative increase in the structural characteristic, evaluated at the mean house price $\overline{P}$ and the mean of the structural characteristic $\overline{X}_n$. The price elasticity equals the coefficient on the structural characteristic times its mean value. For categorical variables the elasticity must be interpreted in a slightly different way. There it indicates the relative increase in the price when a change takes place from one category to another. The price elasticity for categorical variables equals the difference in coefficients between the categories.

The coefficients for the categorical variables have a simple interpretation. They measure the percentage increase in the house price due to a change of dwelling category or in other words the price elasticity of that category. On average a single-family dwelling was 30% more expensive in 1999 than an upstairs apartment. The presence of a lift added 14% (0.07 – (–0.07) * 100%) to the average house price. A change from a three to a four room apartment added 6% ((0.58 – 0.52) * 100%) to the average house price. The elasticities can be converted into monetary values by multiplying them with the average house price in 1999. So the presence of a lift for instance added € 9,230 to an average apartment (0.14 * € 65,932).

A standard robustness check for revealed preference models is to test the forecasting ability. For a subsample of dwellings that were not part of the estimation, the value of the explanatory variables can be filled in the estimated equation. In the case of the hedonic model, an exponential transformation is also needed to obtain forecasted prices. The standard measures to test for the forecasting ability are the Mean Absolute Percentage Error, the Mean Absolute Deviation and the Mean Square Error as given by:

$$MAPE = \frac{1}{n}\sum\frac{|P^i - F^i|}{P^i}$$

$$MAD = \frac{1}{n}\sum|P^i - F^i| \tag{9.7}$$

$$MSe = \frac{1}{n}\sum(P^i - F^i)^2$$

The mean absolute percentage error (MAPE) estimates the absolute difference between the house price (P_i) and its forecasted value (F_i), divided by the house price. A value below 0.1 (or 10%) indicates a very good forecasting ability. Values below 0.3 indicate a decent performance of the model, but when the MAPE exceeds 0.5 the model can be deemed invalid.

The Mean Absolute Deviation (MAD) is the sum of absolute errors (the absolute difference between the actual and predicted value), divided by the number of observations. The Mean Squared Error (MSe) is the sum of the square root of the error terms. It is closely related to the R^2 of the model, since the latter equals 1 minus the MSe divided by the variation in the dependent variable. The root of it is also known as the Root Mean Square error. These measures have little meaning unless there is a frame of reference to compare them with. When for instance two competing models are available, the MAD, MSe and RMSe for the models can be compared. The model with the lowest value for these measures then has the best forecasting ability.

The estimated equation for 1999 is based on a sample of 13,564 dwellings whose value did not exceed € 100,000. It would be interesting to see how well the estimated equation predicts the value of a slightly more expensive subsample of 1,021 dwellings whose value was between € 100,000 and € 110,000. By filling in the values of the structural characteristics and neighborhood dummies in the estimated equation for 1999, the predicted values of the logarithm of the house price for the 1,021 dwellings are obtained. An exponential transformation then yields the predicted house values for this subsample of dwellings.

The Mean Absolute Percentage Error (MAPe) is obtained by averaging the percentage difference between the predicted house values and the original data. For this particular sample of dwellings the MAPe was 0.065. Even though the dwellings were in a higher price range than the dwellings that were used in the estimation, the hedonic house price equation is still able to predict their value with an accuracy of 93.5%. It can be concluded then that the estimated hedonic house price equation for 1999 has high predictive powers. Most statistical software packages have the option to use an equation that was estimated earlier, to predict the values of another sample of observations. As a rule the MAPe and other statistics for the goodness-of-fit of the model will be listed when you carry out the forecasting procedure.

9.6 Discussion

The neo-classical framework for the housing market is explained in detail by Poterba (1984) and DiPasquale and Wheaton (1992). The two models that are highlighted in this chapter reveal the preferences of household categories for items of housing quality and the exchange value of housing attributes on the owner-occupied market. McFadden (1978) is the seminal paper on how discrete choice should be modeled. Discrete choice models for a household's exit choice can, for instance, be found in Bartik et al. (1992) and Knapp et al. (2001). The derivation of the Willingness-To-Pay in an exit choice model is explained in detail by van Ommeren and Koopman (2011).

There are more types of exit choice models in existence than the logit-model for the likelihood to move that is discussed in Sect. 9.4. In duration models the dependent variable is the duration of stay or its derivative function: the hazard rate. The hazard rate is the likelihood that the household moves in a short period of time, given the time it has already spent in the dwelling. From the hazard rate you can deduce how long a household category is expected to stay in its dwelling. Duration models make better use of information that is present in the dependent variable – after all the time spent in a dwelling is continuous and not discrete – and they are superior in their approach to the exit choice, yet more difficult to apply (see Henley 1998).

Estimation of the exit choice is very similar to that of other discrete housing choices: the tenure choice and residential location choice. In the case of the tenure choice, the choice between renting and buying, the household characteristics and the difference between the rent of the rental unit and user costs of the owner-occupied dwelling are the explanatory variables, with the structural characteristics of the dwellings as optional (Mills 1990). In the residential location choice model, the choice of a new dwelling by a recent mover, the structural characteristics, rent or housing costs of the new home enter the model as explanatory variables instead of those of the old home (Quigley 1985).

There are limitations to the revealed preference model for exit and other housing choices (Dieleman 2001 provides a discussion), most of which lie in the interpretation of the estimation results. First the relation that was estimated in Sect. 9.4 is that of the exit behavior of one particular group (a sample of social renters) on one particular housing market (Rotterdam) in one particular period (2002–2004). However, drawing inferences from the outcome of a single regression equation should be done with care. Only re-estimation for different groups at different times can tell if the estimated relation is valid in general for the population for that particular housing market.

Secondly, the low fit of the exit choice model reveals that a lot of choice behavior is left unexplained. This is partly due to missing or unobservable factors. There are, for instance, only three household characteristics in the estimated model. The coefficients on other explanatory variables can conceal the effects of other relevant household characteristics. Care should be given to the interpretation of estimation results, when these secondary effects cannot be controlled for. But even with more household characteristics the exit choice would be hard to explain, because the unobservable traits of the dwelling (e.g. leakage, aesthetic quality), the household (taste, irrationality) and neighborhood (good neighbors) cannot be measured accurately.

In practical applications researchers often use interaction terms between household characteristics and structural characteristics (or neighborhood attributes) in their models. This in effect creates subcategories of households within the categories of dwellings. Although this generally improves the predictive powers of the model, I would strongly argue against this practice unless there are sound theoretical grounds for doing so. The model becomes 'messier' and so will the derivation of odds and marginal effects. And in modeling a relationship a researcher should always strive for simplicity in interpretation and parsimony in the use of explanatory factors.

Thirdly, the regression analysis for the exit choice gives the mean response of household categories to changes in their housing situation. It does not tell you how an individual household in that category responds to these changes, only what the most likely form of action is going to be. You are only allowed to make statements about the average household in that group, not about an individual household. Regression analysis is not the proper method to investigate extremes anyway. Other more qualitative methods are better suited to deal with outlier cases.

The hedonic equation for market rents and house prices that is discussed in Sect. 9.5 is the most common housing market model. The method was introduced by Rosen (1974). It can reveal the contribution of structural characteristics and neighborhood attributes to the rent or price. The hedonic price equation is however just one out of many methods for the valuation and prediction of house prices (Kauko and d'Amato (2008) give an overview of valuation methods). For a large sample of dwellings with little variation in housing quality and prices the hedonic method proves superior. But this superiority is reversed when the dataset becomes smaller and the housing stock is more diverse in its price range or geographical neighborhood (see Borst et al. 2008).

A common fault in the estimation of hedonic equations is the estimation over different market segments. This leads to spatial heterogeneity; structural changes in the price equation across the housing market which are neglected in the estimation of the single equation. Spatial heterogeneity is just one out of many problems that occur when estimation takes place in a spatial context. The spatial econometrics literature (LeSage (1999) is the standard reference) is too vast to discuss here. It deals with many solutions to spatial estimation problems.

The advance of data collection on the micro-scale and GIS-technology allow for better ways to measure the scale of neighborhood attributes and to specify their input: e.g. distance decay functions for accessibility, gravity functions for amenities (Anas 1983) and local averages for socio-demographical attributes (Clapp and Wang 2006). The use of residential survey scores on aspects of residential satisfaction, such as neighborhood satisfaction or place attachment, circumvents many of the measurement problems by using households' direct assessment of the neighborhood (Landale and Guest 1985).

The methodology for measuring neighborhood quality and for spatial estimation is very advanced, and it puts a real strain on the collection of data. Unless you are interested in the impact of a neighborhood attribute on housing choices, area dummies are to be preferred to the use of neighborhood attributes. The inclusion of area dummies is both the simplest and a very effective way to counter spatial estimation problems (Bourassa et al. 2009). Observations that fall inside a neighborhood, postal code area or census tract receive the value one for a particular area dummy and all other observations receive the value zero. Estimation with area-dummies leads to much better estimates of the structural (and household) characteristics, but it is not always easy to interpret the coefficients on the area-dummies themselves.

On a final note: full information, free exchange and the independence of housing choices from past and future choices are very strict assumptions. So strict in fact, that neoclassical models for housing choice for the supply-side (see DiPasquale 1999) and residential location choice models perform rather poorly. This is much more obvious in the estimation of neighborhood attributes than that of structural characteristics. If a house gets bigger, every household understands that the price of it must go up, regardless of the question whether the household in question values the extra space or not. The causal relationship between structural characteristics, housing quality and housing choice and house prices does not need to be questioned.

But when you observe in Table 9.2 that exit rates are lower in neighborhoods with a higher average income, what does that say? Perhaps some households are dissatisfied with the low income of their neighbors and move because of it, even though they cannot observe the income of their neighbors. It might be that an unobservable third factor (quality of shops and schools, crime) is related to neighborhood income and acts as a push-factor. Fewer exits might in themselves cause the higher income of neighbors, when less vacancies lead to excess demand in the neighborhood. Matters get even worse when exit rates and house prices are used as indicators of the stability of the neighborhood by the same households who move out of and into the area. Can you then still say with confidence that you are testing a causal relation?

Research has been conducted that addresses the relation between residential sorting and house prices (Bayer et al. 2005). For all practical purposes I suggest that the coefficients for neighborhood attributes should be treated with more skepticism than the coefficients for structural characteristics. It is also advisable to check on the work of other researchers and see how their estimations compare to yours. If a neighborhood attribute is consistently viewed as a relevant factor in the housing choice, if its coefficient value stays within a certain range, and if there are no questions remaining about the causality of the relationship, only then can you be sure that it reveals the preferences of households for that particular neighborhood attribute.

References

Anas, A. (1983). Discrete choice theory, information theory and the multinomial logit and gravity models. *Transportation Research Part B: Methodological, 17*(1), 13–23.

Bartik, T. J., Butler, J. S., & Liu, J.-T. (1992). Maximum score estimates of the determinants of residential mobility: Implications for the value of residential attachment and neighborhood amenities. *Journal of Urban Economics, 32*, 233–256.

Bayer, P., McMillan, R., & Rueben, K. S. (2005). *An equilibrium model of sorting in an urban housing market*. New Haven: Mimeo, Yale University.

Borst, R. A., Des Rosiers, F., Renigier, M., Stumpf González, M. A., Kauko, T., & d'Amato, M. (2008). Technical comparison of the methods including formal testing of accuracy and other modelling performance using own data sets and multiple regression analysis. In T. Kauko & M. d'Amato (Eds.), *Advances in mass appraisal methods*. Oxford: Blackwell.

Bourassa, S., Cantoni, E., & Hoesli, M. (2009). Predicting house prices with spatial dependence: Impacts of alternative submarket definitions. *Journal of Real Estate Research, 32*(2), 139–160.

Clapp, J. M., & Wang, Y. (2006). Defining neighborhood boundaries: Are census tracts obsolete? *Journal of Urban Economics, 59*(2), 259–284.

Dieleman, F. (2001). Modelling residential mobility: A review of recent trends in research. *Journal of Housing and the Built Environment, 16*, 29–265.

DiPasquale, D. (1999). Why don't we know more about housing supply? *Journal of Real Estate Finance and Economics, 18*(1), 9–23.

DiPasquale, D., & Wheaton, W. C. (1992). The markets for real estate assets and space. *Real Estate Economics, 20*, 181–197.

Henley, A. (1998). Residential mobility, housing equity and the labour market. *The Economic Journal, 108*(447), 414–427.

Kauko, T., & d'Amato, M. (Eds.). (2008). *Advances in mass appraisal methods*. Oxford: Blackwell.

Knapp, T. A., White, N. E., & Clark, D. E. (2001). A nested logit approach to household mobility. *Journal of Regional Science, 41*(1), 1–22.

Landale, N. S., & Guest, A. (1985). Constraints, satisfaction and residential mobility: Speare's model reconsidered. *Demography, 22*(2), 199–222.

LeSage, J.P. (1999). *Spatial econometrics*. Retrieved from: http://www.rri.wvu.edu/WebBook/LeSage/spatial/wbook.pdf.

McFadden, D. (1978). Modeling the choice of residential neighborhood. In A. Karlqvist, L. Lundqvist, F. Snickars, & J. W. Weibull (Eds.), *Spatial interaction theory and planning models* (pp. 75–96). Amsterdam: North Holland.

Mills, E. S. (1990). Housing tenure choice. *Journal of Real Estate Finance and Economics, 3*, 323–331.

Ommeren, J. van & Koopman, M. J. (2011). *Public housing and the value of apartment characteristics to households*. Forthcoming in Regional Science and Urban Economics.

Poterba, J. M. (1984). Tax subsidies to owner-occupied housing. An asset market approach. *Quarterly Journal of Economics, 99*, 729–752.

Quigley, J. M. (1985). Consumer choice of dwelling, neighborhood and public services. *Regional Science and Urban Economics, 15*(1), 41–63.

Rosen, S. (1974). Hedonic prices and implicit markets: Product differentiation in pure competition. *Journal of Political Economy, 82*(1), 34.

Chapter 10
Longitudinal Analysis

Carola de Groot

10.1 Introduction

Research concerning residential mobility and migration is often based on information on either stated intentions to move or on actual moving behavior. Intentions to move are, together with housing preferences (hypothetical housing choices), the subject of the stated preference approach (for an overview see, for example, Kim et al. 2005). The study of actual mobility (as well as actual housing choice) belongs to the revealed preference approach (see Dieleman 1996). This approach is based on observable market behavior, under the assumption that actual behavior is a reflection of preferences (Timmermans et al. 1994).

It is difficult to describe the decision process underlying potential moves by concentrating on either stated intentions to move or on actual moving behavior. Studies using stated preference data are unable to show to what extent stated preferences will result in actual behavior. At the same time, studies using revealed preference data lack the possibility of unraveling the impact of preferences on actual mobility behavior. Actual mobility behavior is not only the result of intentions and preferences; it is the result of an interplay between intentions to move, housing preferences (and the willingness to substitute housing preferences), individual resources and restrictions, and opportunities and constraints at the macro level (compare Timmermans et al. 1994; Goetgeluk and Hooimeijer 2002).

To gain insight into the process between stated intentions to move and actual moves, information is needed on intentions as well as on actual behavior. Information on both intentions to move and actual moving behavior is available in cross-sectional studies as well as in longitudinal studies. Yet, only longitudinal studies combine information about stated intentions and actual moves for the same individuals, by following intended movers in time with respect to their actual

C. de Groot (✉)
PBL Netherlands Environmental Assessment Agency, The Hague, The Netherlands
e-mail: carola.degroot@pbl.nl

S.J.T. Jansen et al. (eds.), *The Measurement and Analysis of Housing Preference and Choice*, DOI 10.1007/978-90-481-8894-9_10,

mobility behavior. The longitudinal research method is, therefore, an ideal methodology to gain detailed insight into both the extent to which people behave according to their stated intentions to move and the circumstances that hamper or stimulate actual moving behavior conditional on the stated intentions to move.

Longitudinal research concerning the question of how intentions to move result in actual moving behavior is often based on small-scale panel surveys in which the same people are interviewed at two or more points in time (Rossi 1955; Landale and Guest 1985; Speare 1974). This can be attributed to the fact that the type of longitudinal data needed to answer the question of how stated intentions to move result in actual moves is difficult to collect. Generally, it is difficult to track the same respondents, especially if the study is conducted over a long period of time. This is particularly problematic in mobility research, because movers are more difficult to trace than non-movers.

Recently, it became possible in the Netherlands to combine survey data with longitudinal register data at the level of individuals. This resulted in a unique large-scale longitudinal data-set that gives information on both intentions to move as well as on actual moving behavior for the same individuals. In contrast to longitudinal studies based on panel surveys, all the respondents can be followed over time; there are no problems with tracking them.

In this chapter, I present analyses that are partly based on the study of De Groot and colleagues (2008), which made use of this newly created longitudinal data-set. The intention to move is derived from the Housing Demand Survey (HDS) 2002; the subsequent moving behavior of HDS respondents is derived from the longitudinal Satellite Spatial and Social Mobility of the Social Statistical Database (SSD) 1999–2005 of Statistics Netherlands (Bakker 2002). I present results of the extent to which stated intentions result in actual moving behavior, and of the circumstances that hamper or stimulate actual moving behavior of those intending to move to another dwelling (the so-called intended *filterers*). Insight into these circumstances can be used to improve the effectiveness of housing market policy.

10.2 Theory and Background

10.2.1 From Stated Intentions to Actual Behavior: Theory

The theory concerning residential mobility and migration starts from the assumption that there has to be a trigger setting off an intention to move. The first source of triggers for moving is the existence of a certain state in the current dwelling or location and the wish to change that state. It is often assumed that the state in the previous dwelling should be dissatisfactory or stressful in order to make people wish to move (Mulder 1996). The second category of triggers for moving relates to the purpose of facilitating the occurrence of life events in the household, educational, or occupational career, such as leaving the parental home, union dissolution, and job change. In contrast to the first category, these triggers for moving are not necessarily related to residential stress or dissatisfaction (Rossi 1955; Clark and Onaka 1985; Mulder 1996.

The trigger for moving partly determines the urgency level of the intention to move. The urgency level of an intention to move indicates how hard people are willing to try, or how much of an effort they are planning to exert, in order to perform the behavior (Ajzen 1991). People who intend to move because the current housing situation does not match with the preferred housing situation (housing reasons) usually do not have an urgent reason to move. Delaying the move only causes an extension of the suboptimal housing situation (Goetgeluk 1997). In contrast, triggers for moving related to life events such as job change, union formation, and especially union dissolution, are more urgent reasons for moving, since these events can hardly take place without at least one move (compare Oskamp 1997; Goetgeluk 1997).

The individual's intention to perform a certain behavior is the central factor in the theory of reasoned action and its later expansion, the theory of planned behavior (Fishbein and Ajzen 1975; Ajzen 1991). These theories have been widely used as models for the prediction of a wide range of behavioral outcomes conditional on intentions (for an overview see, for example, Sheeran 2002). In contrast to the theory of reasoned action (which is designed to predict behavior that is considered to be under volitional control; i.e., persons can decide whether or not to perform the behavior), the theory of planned behavior is designed to predict behavior that is not under volitional control. That is, the execution of certain types of behavior, among which mobility behavior, not only depends on a person's own motivational factors but also on factors that are beyond a person's control (Ajzen 1985, 1991; Madden et al. 1992). Whether people are able to put into effect the intention to move depends on individual resources and restrictions at the micro level, as well as housing opportunities and constraints at the macro level (compare De Jong 1994; Mulder and Hooimeijer 1999).

First of all, the realization of intentions to move is subjected to resources and restrictions. Financial resources are necessary to realize an intention to move (Mulder and Hooimeijer 1999; Kleinhans 2005). The most important financial resource is income. The more income one has, the more dwellings are within the financial reach of an intended mover (Murie 1974; Priemus 1984; Mulder and Hooimeijer 1999). Financial resources are also needed to cover the costs of the move itself. At the same time, restrictions can hamper the realization of an intention to move. Home-ownership may hamper the realization of intentions to move – it requires more time and effort to sell a home than to cancel a tenancy. The household situation is another source of restrictions. It can be more difficult to move for couples and families than for single people, since the former have to take into account the daily activity spaces and preferences of other household members (Mulder 1993; Mulder and Hooimeijer 1999).

Secondly, intentions may not result in behavior if the intention is not strong enough. As a general rule, the stronger the intention to engage in behavior, the more likely that the behavior will be performed (Ajzen 1991). As mentioned before, the urgency of the intention to move is closely connected to the trigger for moving. People whose intention is primarily prompted by housing reasons (which are considered to be less urgent reasons for moving) are therefore expected to be less likely to realize their intention to move than those whose intention is triggered by household or labor market reasons.

The realization of intentions to move probably depends on housing preferences, such as the preference for a rented or an owner-occupied home. While the opportunities in the owner-occupied sector are mainly determined by the affordability of owner-occupied homes (in relation to a person's own financial means and the willingness of the banks to provide mortgage loans), access to the social rental sector is largely dependent on waiting lists and eligibility criteria (such as age and household size). Thus, people who prefer a rental home (which is likely to be a preference for a social rental home since most rental homes in the Netherlands belong to the social housing sector) may have more difficulties realizing their intention to move than those who prefer to move to an owner-occupied home. Tenure preferences may also have an impact on the realization of intentions to move in combination with current housing tenure. Homeowners who prefer moving to a rental home are expected to move more frequently than others because they are considered to have an urgent need to do so (see for example Feijten 2005). For the other combinations of current and preferred tenure, it is not straightforward which combination will lead to more difficulties in finding a new home. On the one hand, homeowners who prefer to move to another owner-occupied home might move less frequently than renters who prefer to move into the owner-occupied sector, because they also have to put effort into selling their home (Helderman et al. 2004). On the other hand, it is also possible that renters have more difficulties realizing their intention to move if they prefer to enter homeownership because of a lack of financial resources or personal equity.

Besides preferences concerning the home, preferences concerning the geographical location are also relevant, since there are great regional differences in housing market opportunities (Clark and Dieleman 1996). The opportunities on the regional housing market are determined by the arrival rate of housing vacancies (Hooimeijer and Oskamp 1996) and are constrained by price and by the allocation rules set by institutions (Hooimeijer and Linde 1988). Compared with the Randstad (the more urbanized western region of the Netherlands), the peripheral regions are characterized by low pressure on the housing market and, therefore, more opportunities to realize an intention to move.

Moreover, the discrepancy between intentions and behavior depends on the stability of intentions between the time of measurement of the intentions and the performance of the behavior (Fishbein and Ajzen 1975). The intention to move may be adjusted, dropped or postponed in time, owing to constraints and facilitators that have been previously ignored by the individual (Lu 1998). For example, if the search for a new home is sufficiently frustrating, a household may decide to stay in the old dwelling (Brown and Moore 1970). The intention to move may also change due to the occurrence of unforeseen life events. During the search for a new dwelling, people may become unemployed or experience other unexpected events such as a divorce or the decease of a household member. Unexpected life events may prevent some people's moves, but may lead others to move rather unexpectedly, even though they might indicate in surveys that they do not intend to move (Hooimeijer and Oskamp 1996). The longer the time interval, the more likely the occurrence of unforeseen events that may change the initial intention to move or to stay in the current home (compare Ajzen and Madden 1986).

10.2.2 From Stated Intentions to Actual Behavior: Results of Previous Research

In previous research, two approaches concerning the link between stated intentions to move and actual moving behavior can be distinguished, both serve the purpose of providing insight into the extent to which intentions to move result in actual moving behavior: the cross-sectional approach and the (internationally more common) longitudinal approach.

10.2.2.1 Cross-sectional Studies

The ideal methodology to analyze the discrepancy between stated intentions and actual behavior is based on a large-scale longitudinal data-set in which people with an intention to move are followed in time with respect to their actual moving behavior. However, until recently, such data were not available in the Netherlands. To get an idea about the extent to which intentions to move result in actual behavior, Goetgeluk and colleagues (1991, 1992) developed a 'quasi-longitudinal' research method using information from a large-scale cross-sectional survey: the cross-sectional Housing Demand Survey. To estimate the extent to which people are likely to realize their intention to move, a comparison was made between the number of people who had moved in the year preceding the moment of interview and the number of people who were actively searching for a home in the period before the moment of interview (see Table 10.1). So, both groups were actively searching for a home in the year before the moment of interview, but only one of them had succeeded in finding a home.

In the last couple of years, several studies have been published that are based on the methodology developed by Goetgeluk and colleagues. However, in some

Table 10.1 Method of cross-sectional analysis

Initial status (t_{i-1})	Actual moving behavior ($t_{i-1} - t_i$)	Current status (t_i)
Actively searching potential mover	Moved	Recent mover
Actively searching potential mover	Not moved	Actively searching potential mover

Source: Based on Goetgeluk et al. (1992)

studies the selection criteria underlying the 'quasi-longitudinal' methodology were less strictly adopted than Goetgeluk and colleagues had recommended: the selection of respondents intending to move comprised not only those actively searching for a home, but a wider selection of intended movers (for example see Van Groenigen and Van der Veer 2006; VROM 2007a; Haffner et al. 2008).

The study of Goetgeluk and colleagues estimated that of the 4,222 respondents who were searching for a home in the year before the moment of interview, about 58% moved to another home, while the remaining respondents were still searching for a home (Goetgeluk et al. 1992). Using data from a survey held in 1994, Hooimeijer and Poulus (1995) estimated that in the year preceding the moment of interview, about 416,000 people changed residence. In the same period about 461,000 people had been searching for a home without success. So, of the 877,000 respondents who have been actively searching for a home in the year preceding the moment of interview, only 416,000 had found a new home. This results in an estimated rate of success in finding a home within 1 year of 47% (416,000/877,000). The cross-sectional analysis of Van Groenigen and Van der Veer (2006) for Amsterdam shows that in 2005 71,000 households stated that they definitely intended to find a home in Amsterdam within 2 years, while about 39,100 households had recently moved (in the 18 months before the moment of interview). By relating these numbers to each other, Van Groenigen and Van der Veer (2006) found a success rate of 55% in 2005. The study of Haffner and colleagues (2008) for the Netherlands as a whole shows a somewhat lower success rate of 47%. A national cross-sectional analysis, again in the Netherlands, by VROM (Dutch Ministry of Housing, Spatial Planning and the Environment) (2007a) estimated that in 2006, about 1.9 million people declared an intention to move within 2 years, whereas in the last 2 years only 1.1 million people had moved. The success rate of 58% suggests that a considerable proportion of intended movers will probably not realize their intention. According to VROM (2007b), the rate of success is especially low among people aged below 25: the number of people who are searching for a home far exceeds the number of people who have recently moved. Another analysis by VROM (2007c), in which a comparison is made between the number of people who have recently moved (1.1 million) and the number of people with an urgent intention to move (1.1 million), suggests that most people with an urgent intention to move will find a home within 1 year.

Studies using cross-sectional data give an idea about the extent to which intentions to move are likely to result in actual moves. However, one should bear in mind that estimations of success rates derived from confrontations between aggregated data about stated intentions and aggregated data about actual moves are not without problems. The success rate might be overestimated because the actual moves might include 'sudden' moves of people who initially had no intention of changing residence (for example, people who need to move unexpectedly due to an unforeseen divorce). But, the success rate might also be underestimated because some people with an intention to move might not be in a hurry and might, therefore, spend quite some time finding a new home. The actual probability of moving among those with an intention to move can only be determined using longitudinal data in which people with an intention to move are followed in time with respect to their actual moving behavior.

10.2.2.2 Longitudinal Studies

Longitudinal studies, often based on panel data, also show a discrepancy between stated intentions to move and actual moving behavior. Using panel data for Philadelphia, Rossi (1955) found that of those who expected to move, about 80% did so within the following year. Two decades later, Speare (1974) concluded that about 37% of those with a wish to move in Rhode Island actually did so. Landale and Guest (1985) concluded that of the residents of the Seattle metropolitan area thinking about moving, almost 40% moved within 1 year. Konter and Van den Booren (1988) used a panel survey for the province of Noord-Holland in the Netherlands, and showed that about 28% of those with a desire to move changed residence within 1 year. A small-scale panel survey for the Netherlands by Van Kempen and colleagues (1990) showed that about 15% of those living in the city of Utrecht and intending to move changed residence within 1 year. Based on a small-scale panel survey for Utrecht and Arnhem (relatively large cities within the Netherlands), Goetgeluk (1997) found that, of those actively searching for a home, about 50% realized their intention to move within 1 year. Lu (1998) used the national American Housing Survey (AHS) and showed that 44% of those preferring to move realized this preference within 2 years. Based on the Panel Study of Income Dynamics (PSID), a representative survey of U.S. residents and their families, Kan (1999) showed that almost half of the households expecting to move did so within the following 2 years. Recently, a study by GfK (2009) showed that about 42% of those with an intention to move in the cross-sectional Housing Research Netherlands survey 2006 changed residence within 2 years; in this study, the moving behavior in the years after the moment of interview was determined from the municipal population registers.

All the above-mentioned longitudinal studies show a discrepancy between inten tions and behavior, but the size of the discrepancy differs considerably between the studies. These differences are probably partly caused by the diversity of the intention concepts used in the studies. For example, Speare (1974) measured mobility 'wishes', while Kan (1999) used mobility 'expectations'. Generally, it is assumed that expectations are a better predictor of moving behavior than desires or wishes to move (Rossi 1955; Lu 1998; Crowder 2001; Sheeran 2002). When individuals form a plan or expectation to move, it is likely that certain constraints on migration have been taken into account (Lu 1998). Therefore, mobility expectations reflect what is perceived as likely over a specified time period, whereas desires, wishes and thoughts represent, to a greater extent, unconstrained preferences (Crowder 2001).

Longitudinal studies also found evidence that some people move although they previously had not intended to do so (Rossi 1955; Landale and Guest 1985; Lu 1999; Kan 1999). Rossi (1955) and Kan (1999) showed that unforeseen life events are important triggers of such not previously intended moves. Unforeseen life events can cause an urgent reason for moving and a subsequent move within a short period of time (compare Speare 1974).

In contrast to cross-sectional studies, longitudinal studies can also provide more insight into the extent to which circumstances hamper or stimulate actual moving behavior conditional on a positive attitude to move. Such studies show that the realization of stated intentions to move depends on individual and household characteristics. A study for the Netherlands, based on a follow-up survey among 4,000 respondents of the Housing Demand Survey (HDS) 1985/1986, indicated a strong linkage between age and the extent to which people realize their intention to move (Everaers and Lamain 1989). Other studies have shown that whites are more likely to realize their intention or expectation to move than blacks (Lu 1999; Kan 1999; Crowder 2001). The impact of income on the realization of stated intentions to move is not straightforward. Some studies have shown a positive effect of income on the probability of moving for those intending or expecting to move (Moore 1986; Everaers and Lamain 1989; Lu 1998), although Everaers and Lamain found that this only applies to people aged below 45 years, while other studies hardly found an income effect (Goetgeluk 1997; Kan 1999). Furthermore, living in a rented or crowded home increases the probability of moving for those intending or expecting to move (Moore 1986; Lu 1998; Duncan and Newman 1976). Finally, some studies showed that the probability of moving among those intending to move differs between regions (Lu 1998; Kearns and Parkes 2003).

10.3 Data and Methods

10.3.1 Longitudinal Data

In order to gain insight into the process between stated intentions and actual moving behavior, a researcher needs longitudinal data about the stated intentions and actual moving behavior of the same individuals. Most longitudinal studies in which the realization of intentions to move is analyzed are based on panel data (Rossi 1955; Landale and Guest 1985; Speare 1974). Data on moving behavior is obtained in a follow-up survey, from the same respondents. Panel surveys are often faced with difficulties in tracking the same respondents in time because of the drop out of respondents caused by moves, deaths or refusals to continue (for more information on panel attrition see Ruspini 1999; Behr et al. 2005). In contrast to the above-mentioned studies, the longitudinal studies of Lu (1998), (1999) and Kearns and Parkes (2003) are based on surveys that follow housing units in time rather than the households that occupy them: the American Housing Survey (AHS) and the English House Condition Survey (EHCS). Changes in the occupants of any housing unit between successive surveys are interpreted as actual moves of the initial occupant of the housing unit. Unfortunately, changes can also occur if the initial occupant has died in the period between the successive surveys, but this information is not always available, as is the case in the data-set of Kearns and Parkes (2003).

Recently, it became possible in the Netherlands to combine survey data with longitudinal register data at the level of individuals. These two data sources can be linked by using a unique personal identification number; this resulted in a unique longitudinal data-set that is similar to a linked panel. In linked panels, data items that are not collected primarily for panel purposes (such as census or register data) are linked together using unique personal identifiers (Ruspini 1999). In contrast to panel surveys, there are no problems with tracking the respondents in time. In fact, linking data items by using a personal identifier is the least intrusive method of collecting longitudinal data (Buck et al. 1994). In the rest of this section, a detailed description is given of the design of the longitudinal data-set used by De Groot and colleagues (2008), in which information about intentions to move (survey data) is enriched with information about actual moves (register data) for the same individuals.

10.3.1.1 The Enriched Housing Demand Survey

Since 1964, large surveys have been held in the Netherlands to gain insight into the housing situation of individuals and households. These Housing Demand Surveys (HDS) and its 2006 successor the Housing Research Netherlands survey (HRN) are conducted every 4 years in the Netherlands. The surveys are based on large person-based samples that represent the Netherlands population aged 18 and over and who are not living in institutions. The fieldwork of the surveys differs from 6 months to 1.5 years; the interviews of the HDS 2002 – the survey used in our study – were held in the period January 2002 until March 2003. The surveys contain detailed information about socio-demographic and socio-economic characteristics, the current housing situation, the intention to move, and preferences concerning the future home and the residential location. Retrospective information about the moving behavior of individuals in the 2 years before the interview is also included.

For the purpose of this research, the HDS 2002 data were enriched with register data from the longitudinal Satellite Spatial and Social Mobility of the Social Statistical Database (SSD) of Statistics Netherlands (for more information see Bakker 2002). To link the HDS 2002 with the SSD data on the level of individuals, a unique personal identification number is used. It should be mentioned, that it is also possible to link other survey years with SSD data as long as the data contain personal identification numbers issued by Statistics Netherlands that can be traced back to the population register. At the moment, this is the case for the HDS 1998 and 2002, and for the HRN 2006 survey. In order to follow the respondents in time, the HDS 2002 was linked with SSD years 2001–2002, 2002–2003, 2003–2004 and 2004–2005 (which was the most recent SSD year available at the time of the research).

The core of the SSD consists of linked registers, containing administrative demographic and socio-economic data for all persons who have been registered as living in the Netherlands at any moment since 1995 for the period 1999–2005. Within the SSD, several SSD Satellites have been developed, including the Satellite Spatial and Social Mobility. This Satellite contains information about household and socio-economic characteristics such as household composition and employment status, housing tenure and the value of the property used for tax purposes as well as the moving behavior of individuals. Information about moves and other individual characteristics refer to the situation (or changes in this situation) between the start of the SSD year (usually the last Friday of September in year t_i) and the end of the SSD year (the last Friday of September in year t_{i+1}). Information about moves originates from the municipal population registers (in Dutch: Gemeentelijke Basis Administratie, abbreviated GBA). A disadvantage of using register data to determine the moving behavior of individuals is that people do not necessarily live at the address where they are registered. This limitation probably results in somewhat biased estimates. The magnitude of this problem is unknown, but it is probably most severe for students and other mobile groups.

To determine the extent to which people's behavior conforms to their stated intention to move, the moving behavior of HDS respondents is determined for a period of 2 years after the moment of interview. For the HDS 2002, this implies that those interviewed in January 2002 were followed up to January 2004, whereas those interviewed in March 2003 were followed up to March 2005. The exact date of the first move after the moment of interview (year month format) is not included in the Satellite, but has been directly derived from the GBA. A move (which covers migration within the Netherlands as well as emigration) was measured as the first relocation after the interview; further relocations were ignored.

To determine whether respondents have died in the 2 years after the time of the HDS 2002 interview, information about the exact date of death of respondents is necessary. This information was obtained from the GBA and was linked to the longitudinal data-set by using the personal identification number.

10.3.1.2 Accessibility of Data

Access to SSD data and to HDS data including the personal identification number of the HDS respondents is provided by Statistics Netherlands. Statistics Netherlands permits institutions such as universities and planning agencies to work with micro data under very strict conditions. Because the data comprise a very detailed picture of every inhabitant of the Netherlands, data security and privacy are important issues. There are legal conditions that prevent the disclosure of individual data by Statistics Netherlands. Access to the data is arranged by on-site constructions, sometimes accompanied by cooperation between Statistics Netherlands and the institute involved. For reasons of confidentiality and of data protection legislation, the generated output is checked before it is released to the researcher (for more information see Statistics Netherlands 2006; Gouweleeuw 2006).

10.3.1.3 Selection of Respondents

The HDS 2002 contains information on 74,788 respondents. For the analyses in which the HDS respondents are followed in time, a small number of HDS respondents who died in the 2 years after the time of interview were excluded. After this selection, the research sample included 73,714 respondents. The majority of the analyses refer to the respondents who expressed an intention to move. In the HDS, intentions to move are measured by the question: *"Do you want to move within the next 2 years?"* (In Dutch: *"Wilt u binnen twee jaar verhuizen?"*). If someone answered *"Possibly yes, maybe"*, *"I would like to, but I cannot find anything"* or *"Most certainly yes"*, this person is considered to have an intention to move. If someone answered *"Most certainly not"* or *"I do not know"*, this person is considered not to have an intention to move. In the research sample, 18,143 respondents (25%) expressed an intention to move, 53,803 respondents (73%) expressed no intention to move, and 1,768 respondents (2%) expected an involuntary move (for example because the dwelling was going to be demolished) or had already found a new home.

Among the intended movers, a distinction was made between starters and filterers; two different groups on the housing market. Starters are people who expressed an intention to move to their first independent dwelling; they want to leave the parental home or want to move out of their dormitory in order to establish their own household. Filterers are people who expressed an intention to move from one independent housing situation to another. Of all the respondents with an intention to move, 4,543 respondents are classified as intended *starters* (25%) and 11,730 respondents as intended *filterers* (65%). The remaining 1,870 respondents include people who intend to leave the housing market (because they want to move abroad or into a nursing home) and people who intend to move into a dormitory.

10.3.2 Variables

The analyses are partly based on the analyses of De Groot and colleagues (2008). With the exception of the dependent variable, all variables are derived from the HDS and refer to the moment of the HDS 2002 interview.

The dependent variable reflects whether people moved in the 2 years after the interview was held, and was coded into two categories: moved (1) and did not move (0).

The multivariate analysis for filterers (see next section) includes several independent variables. The descriptive statistics and definitions of the dependent and independent variables used in the multivariate analysis are presented in Appendix 1. The appendix also contains the percentage of movers in the distinct categories of the independent variables.

The urgency of the intention to move is based on the urgency variable in the HDS. Respondents with an urgent intention to move are intended movers who have indicated that they want to move within a year, who are actively searching for a new home (they have undertaken at least one search activity and the reason why they have not moved yet cannot be traced back to a low search activity) and who have declared that they will react immediately if they find a desired home. Intended movers who do not meet these requirements are considered to have a less urgent intention to move; most of the filterers in our research sample who stated an intention to move have a low urgency for moving (79%).

The reason for moving refers to the main motive for moving and was coded into four categories. The first category covers demographic reasons such as marriage or cohabitation, union dissolution (either married or unmarried), and aging and health. The other categories refer to reasons related to the home or the neighborhood characteristics (housing reasons), reasons related to education or work, and to 'other' (highly diverse) reasons for moving.

Household income was coded in quartiles; the quartiles were based on the household income of all respondents in the survey sample. Income was based on household income and not personal income because mobility behavior is likely to be based on the total income in the household (it should be noted, however, that about 15% of the intended filterers expect a change in their household situation, which might affect current household income).

Level of education was based on the highest achieved educational level and was categorized into three levels: up to lower secondary or unknown; higher secondary or medium vocational training; and higher vocational training or university. Those whose educational level is unknown (N=99) were assigned to the category 'up to lower secondary'.

The expected household composition was classified using prospective information about the expected household situation after the intended move. It was coded into three categories, singles (one-person households, and persons living with at least one other person at an address with no clear relationship to the other person(s), couples without children, and families with children (couples with children and one-parent families).

The preferred location was divided into three search regions within the Netherlands: the Randstad (core region), the surrounding intermediate zone, and the periphery of the Netherlands. For the classification, I used a gravity equation on total population size in all municipalities (496 in total) at the beginning of 2002. Based on the gravity value of the municipality, all 496 municipalities were assigned to a region (for more information about the methodology see Van Oort 2003). In order to create uninterrupted zones, there were 11 municipalities where the classification differed from the classification of the surrounding municipalities. These were then assigned to the surrounding region. Based on the preferred place of residence, the preferred location could be specified for most respondents. For those (N=2159) who did not state a preferred place of residence, information about the preferred province has been used. Because ten provinces (of the twelve provinces in total) are roughly situated within one single search region, I could specify that a preference for province X_i (i=1,..., 10), was similar to a preference for region N_i (i=1,...,3). The two other provinces (Noord-Holland and Flevoland) are spread over all three search regions. Those who expressed a preference for one of these provinces (N=333) were assigned to the search region in which the current place of residence was located. Because the majority (about 75%) of those who stated a preference for one of these provinces were already living in the preferred province – and most people move short distances (Clark and Dieleman 1996) – the current place of residence represents the most plausible search region for most respondents. For the small number of respondents who were currently not living in the preferred province, the chosen criterion leads to a rather arbitrary assignment to a search region.

The perceived health situation was measured using information about the respondents' opinion about their own health situation, and was coded into three categories: healthy (which covers the HDS categories: *"Good"*; and *"Very good"*); reasonably healthy; and less healthy (which covers the HDS categories: *"Sometimes good, sometimes bad"*; and *"Bad"*).

Crowding was calculated by dividing the number of rooms by the number of persons in the household. It was categorized in three levels: neutral (between one and two rooms per person in the household), crowded (one room or less per person in the household) and spacious (two or more rooms per person in the household). The classification of crowding – less than one room per person – is in line with the standard commonly used (Myers et al. 1996, see also Gray 2001).

The degree of satisfaction with the home and the neighborhood were both coded into three categories: satisfied (which refers to *"Satisfied"*; and *"Very satisfied"*); neither satisfied nor dissatisfied; and dissatisfied (which refers to *"Dissatisfied"*; *"Very dissatisfied"*).

10.3.3 Method

The descriptive results start with a bivariate, longitudinal analysis to investigate the extent to which people behave according to their initial stated intention to move or to stay. Among the intended movers, a distinction was made between starters and filterers. Furthermore, a cross-sectional comparison was presented between the number of people who changed residence in the year preceding the moment of interview and the number of people who were supposedly searching for a home in the year preceding the moment of interview. The descriptive results section ends with a comparison between the longitudinal research method and the cross-sectional research method. In order to correct for selective non-response, the HDS sampling weights are used in the descriptive analyses. Most surveys are confronted with selective non-response; ethnic minority groups, for example, often have lower response rates than natives (Feskens et al. 2006). This non-response may lead to serious bias in the results concerning the discrepancy between stated intentions and actual mobility behavior, especially if the non-response group is considerable in size. For example, if natives are overrepresented in the survey, and the realization of stated intentions to move is selective with respect to ethnicity, the extent to which intentions to move result in actual moving behavior might be over- or underestimated. Since it is the respondent who is tracked in time, and not the complete household, the sampling weights of persons are used. These weights are based on age, gender, marital status, country of origin, and household characteristics (Meeuwissen et al. 2003).

After the descriptive analyses, a multivariate analysis was performed in order to gain insight into the circumstances that hamper or stimulate the actual moving behavior of those intending to move. This analysis was restricted to those intending to move to another dwelling (that is, for intended filterers). A logistic regression model was used because of the dichotomous character of the dependent variable. A logistic regression estimates the odds of a certain event occurring. If *P* is the probability of the occurrence of an event, and *O* is the odds of the event, then the odds is the probability of the event occurring divided by the probability of no event: $O = P/(1\text{-}P)$. Logistic regression applies the maximum likelihood estimation after transforming the dependent variable into a logit (the natural log of the odds of the event occurring or not). The logistic regression model for *k* independent – explanatory – variables (*X*) is formally specified as follows (Allison 1999):

$$\log\left[\frac{p_i}{1-p_i}\right] = \alpha + \beta_1 X_1 + \beta_2 X_2 + \beta_3 (X_1 X_2) + \cdots + \beta_k X_k, \qquad (10.1)$$

where α is the intercept (or 'constant term'), and $\beta_1, \ldots, \beta_k$ are the logistic regression coefficients. The equation given above also includes an interaction term $(X_1 X_2)$; this term may be added to the model in order to examine whether the effect of X_1 on the dependent variable varies with values of X_2. Or, as Garson (2009) puts it, an interaction term incorporates the joint effect of two independent variables on a dependent variable over and above their separate effects (for more information about interaction terms, see Jaccard 2001).

The impact of the independent variables is frequently reported in terms of odds ratios. The odds ratio is the exponent (antilogarithm) of β. The odds ratio for continuous independent variables represents the factor by which the odds of the event change for a one-unit change in the independent variable; for categorical independent variables, it estimates how much more likely (or unlikely) it is for the outcome to be present among those belonging to one particular category than among those belonging to the reference category of the independent variable. To illustrate, the odds ratio of the urgency of the intention to move (presented in Table 10.2) estimates that a move is 3.7 times more likely to occur among those with an urgent intention to move than among those with a less urgent intention to move (for more information on logistic regression analysis and the interpretation of estimated parameters see, for example, Hosmer and Lemeshow 1989).

To test the fit of the logistic regression model, a likelihood ratio test can be performed. This test tests the significance of the difference (Chi^2) in the likelihood ratio (−2 times the log likelihood) of the model with all explanatory variables and the model with only an intercept (and no explanatory variables). A well-fitting model (which implies that the full model is significantly different from the reduced model)

Table 10.2 Logistic regression of moving (ref: not moving) for intended filterers

	B	S.E.		Exp(B)
Characteristics of intention to move and housing preferences				
Urgent intention (ref: less urgent intention)	1.300	0.051	***	3.668
Reason for moving (ref: demographic event)				
Housing	−0.256	0.069	***	0.774
Work or education	−0.118	0.100		0.888
Other reason	−0.150	0.074	**	0.860
Wish to own (ref: wish to rent)	0.176	0.065	***	1.193
Wish to own, homeowner (ref: wish to rent, renter)	−0.365	0.114	***	0.694
Wish for single family home (ref: wish for apartment)	−0.034	0.057		0.966
Preferred search location (ref: national periphery)				
Randstad	−0.191	0.056	***	0.826
Intermediate zone	−0.120	0.057	**	0.886
Resources and restrictions				
Income (ref: lowest quartile)				
Second quartile	−0.102	0.065		0.903
Third quartile	0.066	0.074		1.069
Highest quartile	0.103	0.084		1.109
Level of education (ref: up to lower secondary)				
Higher secondary or medium vocational	0.030	0.055		1.030
Higher vocational or university	0.105	0.059	*	1.111
Not employed (ref: employed)	−0.044	0.061		0.957
Homeowner (ref: renter)	0.357	0.102	***	1.429
Expected household composition (ref: single)				
Couple without children	0.144	0.066	**	1.155
Family with children	−0.141	0.079	*	0.869
Other individual and housing characteristics				
Age (ref: <25 years)				
25–34	−0.188	0.079	**	0.828
35–44	−0.595	0.086	***	0.552
45–54	−0.780	0.095	***	0.458
55–64	−0.872	0.110	***	0.418
≥65	−0.539	0.117	***	0.583
Ethnicity (ref: native-born)				
Non-western immigrant	−0.286	0.073	***	0.752
Western immigrant	−0.082	0.075		0.922
Perceived health (ref: healthy)				
Reasonably healthy	−0.202	0.080	**	0.817
Less healthy	−0.186	0.087	**	0.830
Single family home (ref: apartment)	−0.209	0.053	***	0.811
Crowding (ref: neutral)				
Crowded	−0.033	0.063		0.968
Spacious	−0.212	0.060	***	0.809

(continued)

Table 10.2 (continued)

	B	S.E.		Exp(B)
Satisfaction with home (ref: satisfied)				
Neither satisfied nor dissatisfied	0.007	0.061		1.007
Dissatisfied	0.182	0.072	**	1.200
Satisfaction with neighborhood (ref: satisfied)				
Neither satisfied nor dissatisfied	–0.097	0.062		0.907
Dissatisfied	–0.040	0.058		0.961
Constant	–0.174	0.134		0.840
Initial -2 log likelihood	14,641			
Model -2 log likelihood	13,490			
Improvement (Chi²)	1,151; df=34; p=0.000			
Nagelkerke R^2	0,131			
N	11,730			

Source: SSD Satellite Spatial and Social Mobility 1999–2005, including HDS 2002
Note: 95% confidence intervals can be calculated as: parameter (*B*) minus 1.96 * S.E. (lower boundary), and parameter plus 1.96 * S.E. (upper boundary)
*p<0.10; **p<0.05; ***p<0.01

is a model in which at least one of the explanatory variables is significantly related to the dependent variable (Garson 2009, for more information, see also Yamaguchi 1991). The likelihood ratio test can also be used to test the impact of interaction effects; in this case, the difference is tested between a full model and a model in which the interaction effect has been dropped.

The logistic regression model was estimated without using sampling weights, since these weights are roughly a function of independent variables included in the logistic regression model (and therefore the selective non-response is directly controlled for). In such cases, unweighted regression estimates are preferred because they are unbiased, consistent, and have smaller standard errors than weighted regression estimates (Winship and Radbill 1994).

10.4 Stated Intentions to Move and Actual Moving Behavior: Results

10.4.1 Descriptive Results

10.4.1.1 Longitudinal Analysis

In 2002, 23% of all HDS respondents stated an intention to move within the next 2 years. After 2 years, about one third (31%) had realized this intention to move. It is often assumed that especially intended *starters* (people who intend to move to their first independent dwelling) have a difficult position on the housing market (see for example Kruythoff 1994; REA 2006). Nevertheless, if stated intentions are compared with actual moving behavior, it becomes clear that starters move more frequently than intended *filterers* (people who stated an intention to move from one independent housing situation to another). Of the starters in 2002, almost 44% realized their intention to move within 2 years, compared with 31% of the filterers in 2002. Apparently, there is a large gap between stated intentions to move and actual behavior; the majority of individuals are unable to realize their intention within 2 years, or might postpone or give up their initial intention to move. This holds for starters as well as for filterers.

At the same time, a small proportion – about 6% – of those who did not intend to move in 2002 moved within 2 years. According to several researchers (see, for example, Rossi 1955; Kan 1999), unforeseen life events can trigger people to move, although they did not have any intention to change residence before the life event took place. These movers did not only develop an intention to move after the time of interview, but they also succeeded in realizing this intention in a relatively short period of time.

The longitudinal analysis shows a substantial gap between intentions and actual behavior; most people with an intention to move do not realize this intention within 2 years, and some people move even though they did not state an intention to do so at the time of the interview. The results are in line with other longitudinal studies, which generally show a relatively large discrepancy between stated positive attitudes towards moving (e.g. desires or intentions) to move and actual moving behavior. At the same time, the results show that intended movers relocate considerably more frequently than those with no intention to move. So, despite the large gap between stated intentions and actual behavior, intentions to move are important – although not perfect – predictors of actual moving behavior (compare Lu 1998; Parkes and Kearns 2003).

10.4.1.2 Longitudinal Versus Cross-sectional Approaches

Not only longitudinal studies, but also cross-sectional studies show a discrepancy between stated intentions to move and actual moving behavior (see Sect. 10.2.2). But a comparison between the two approaches shows that the cross-sectional approach results in a different picture than the longitudinal approach (see Fig. 10.1).

For the cross-sectional comparison it is calculated that about 1.1 million people changed residence in the year preceding the time of interview, while about one million people were searching for a new residence for at least the 6 months prior to the time of interview. Of the total of approximately 2.1 million people who were supposedly searching for a home in the year preceding the moment of interview (the people with an 'inferred' intention to move in Fig. 10.1), about 1.1 million found a new residence; this results in an estimated rate of success of 53%. However, the longitudinal analysis shows that of those who stated an intention to move within *1 year* (third column in Fig. 10.1), only 32% moved within this period (note that the period differs from the period in the analysis presented in the

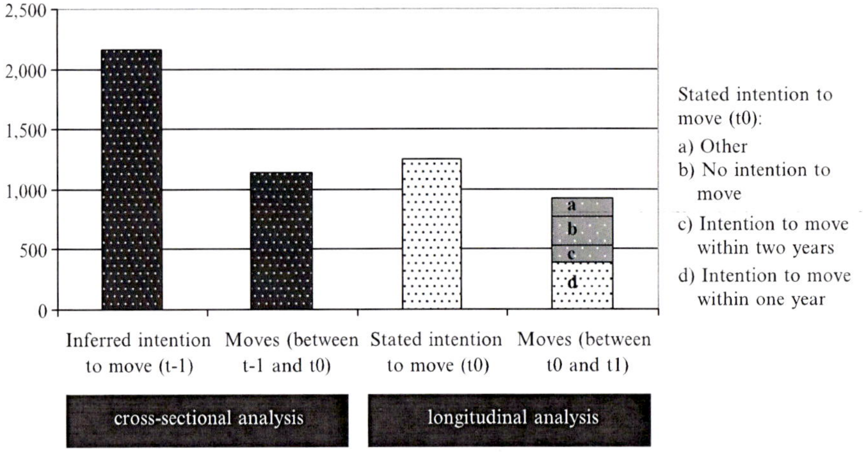

Fig. 10.1 The discrepancy between intentions and moving behavior for a 1 year period: cross-sectional versus longitudinal approach (×1,000), weighted data. Note: The category 'other' refers to the movers who expected an involuntary move (for example because of housing demolition or renovation), or who had not moved yet but had already found a new home at the moment of interview (*Source*: SSD Satellite Spatial and Social Mobility 1999–2005, including HDS 2002)

previous section in order to make a proper comparison between the cross-sectional and longitudinal method).

The cross-sectional analysis seems to result in a smaller discrepancy between stated intentions and actual behavior than the longitudinal analysis. This can be explained by the fact that the total number of actual moves in the year before the year of interview far exceeds the total number of moves made by people with a stated intention to move within 1 year. The longitudinal analysis makes clear that this is most likely related to the fact that a substantial part of the actual movers in a certain period had no stated intention to move at the start of the period. So, the discrepancy between stated intentions to move and actual moving behavior is much larger on the level of individuals than on the aggregate level.

The cross-sectional approach does not only result in a smaller discrepancy between stated intentions and actual behavior, but may also lead to other conclusions. A cross-sectional analysis of VROM (2007b) shows that the rate of success is especially low among people aged below 25: the number of people intending to move exceeds by far the number of people who have recently moved (see also Sect. 10.2.2). In contrast, most longitudinal studies show that people in this age group realize their intention to move more often than those in older age groups (see for example Lu 1999). Furthermore, a cross-sectional analysis in which the number of people with an urgent intention to move is compared with the number of people who have recently moved suggests that most people will find a home within 1 year (VROM 2007c). In contrast, the longitudinal study of De Groot and colleagues (2008) shows that even people with an urgent intention to move often do not realize this intention within the stated period.

10.4.2 Multivariate Results of the Longitudinal Analysis

The logistic regression model shows the extent to which the relatively low realization of intentions to move among filterers can be ascribed to the urgency of the intention to move, housing preferences, and resources and restrictions (see Table 10.2).

10.4.2.1 The Influence of Characteristics of the Intention to Move and Housing Preferences

One explanation as to why so many filterers do not realize their intention to move can be found in the urgency of the intention to move. First of all, most filterers (79%) have a less urgent intention to move. Those with an urgent intention to move are estimated to be 3.7 ($e^{1.300} = 3.668$) times more likely to move as those with a less urgent intention to move. This is in line with the general rule that the stronger the intention is to engage in behavior, the more likely it is that the intention is realized (Ajzen 1991). Secondly, most filterers intend to move for housing reasons (De Groot et al. 2008). As expected, those intending to move because of housing have a lower probability of realizing their intention than those who want to move because of demographic events.

Housing preferences have a significant effect on the moving behavior of filterers. The extent to which filterers realize their intention to move depends on tenure preferences, but only in combination with the current tenure. Although the *Nagelkerke* R^2 hardly changes after including the interaction term, the likelihood ratio test showed that the interaction effect contributes significantly to the multivariate model ($Chi^2 = 10$; $df = 1$; $p = 0.001$) and should therefore be retained. The main effect of having a preference for an owner-occupied home, as reported in Table 10.2, should be interpreted as the effect for current renters (since the interaction variable relates tenure preferences and current tenure, and 'renter' is the reference category in the current tenure variable). The main effect shows that current renters with a preference to move to an owner-occupied home are 1.2 times (see Table 10.2; $e^{0.176}$) more likely to move within 2 years than renters who intend to move within the rental sector (from rent to rent). The total effect of the main effects and interaction effect is illustrated in Fig. 10.2. Homeowners with a preference to move to a rental home are estimated to be 1.4 times more likely to move within 2 years than renters who prefer to move to another rental home.[1] Renters who want to move within the rental sector of the housing market have in fact the lowest probability of moving. This might be an indication of barriers in

[1]The total effect is the summation of the main effects and the interaction effect. For example, the total effect for homeowners (1) with a preference to move to a rental home (0): $0.357*(1) + 0.176*(0) - 0.365*(1*0) = 0.357$; the corresponding predicted odds ratio is 1.429 ($e^{0.357}$).

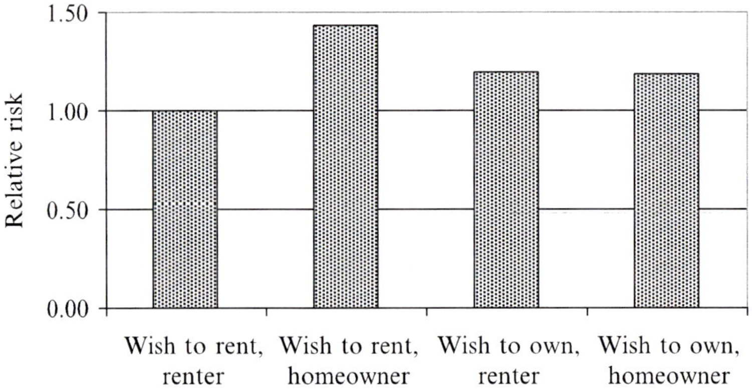

Fig. 10.2 Relative risk of moving for those preferring a rental home or an owner-occupied home by current tenure (based on the total effect of the main effects and interaction effect of 'tenure preferences' and 'current tenure' in Table 10.2) (*Source*: SSD Satellite Spatial and Social Mobility 1999–2005, including HDS 2002)

the social rental sector. Figure 10.2 also shows that homeowners with a preference for moving to a rental home are more likely to move within 2 years than homeowners who prefer to move to another owner-occupied home. This can be explained by the fact that homeowners usually do not move out of owner-occupied dwellings unless there is an urgent need to do so. Finally, among those preferring to move to an owner-occupied home, renters are as likely as homeowners to realize their intention to move within 2 years. At first sight, this seems to contradict the general expectation that it is relatively difficult to move to an owner-occupied home for those who do not yet own a home (because of a lack of financial means or personal equity). However, one should bear in mind that the model offers only an explanation for the extent to which people realize their intention to move, and not for the extent to which people succeeded in moving to an owner-occupied home.

Filterers who prefer a single family home do not have a significantly lower probability of realizing their intention to move than those preferring an apartment. Although people who prefer a single family home might have fewer opportunities to realize their intention to move (because in the Netherlands such homes are relatively more scarce on the housing market than apartments; see for example VROM 2003, 2007a), this is not reflected in the probability of moving.

Furthermore, preferences concerning the search location play a significant role in the extent to which one acquires a new home within 2 years time. As expected, filterers who stated a preference for a home in the Randstad – an area where the housing market is under great pressure – are 0.8 times as likely to move as those who intend to search for a home in the national periphery, an area where the pressure on the housing market is lower. Those who prefer to search for a home in the intermediate zone also have a lower probability of moving.

10.4.2.2 The Influence of Resources and Restrictions

Somewhat surprisingly, financial resources – income, level of education and the employment situation – have hardly any effect on the moving behavior of filterers. Only filterers with higher vocational or university education are 1.1 times more likely to move than those with up to lower secondary education, but this effect is only marginally significant. The absence of a significant income effect may indicate that less wealthy people prefer to move to other (more affordable) types of homes than wealthy people. Moreover, the finding does not imply that financial resources are completely irrelevant in the mobility decision process. As De Groot and colleagues (2008) have shown, financial resources are taken into account when people state an intention to move to another dwelling. When people believe that they have little control over fulfilling this intention because of a lack of resources or opportunities, their intentions to carry out the behavior may be low (compare Madden et al. 1992).

The realization of intentions to move among filterers is restricted by the household situation. As expected, singles have a greater probability of realizing an intention to move than families. Conversely, their probability of moving is lower than that of couples. This cannot be attributed to the fact that singles often have lower incomes than couples, since income differences were controlled for. It is possible that singles tend to search in housing sectors in which the availability and affordability of dwellings is problematic.

Furthermore, the current housing tenure situation plays a role (see also previous section). The main effect of current tenure in the model should be interpreted as the effect for those who prefer to move to a rental home (since the interaction variable relates current tenure with tenure preferences, and 'preference to rent' is the reference category in the tenure preference variable). The main effect shows that homeowners who intend to move to a rental home are 1.4 times more likely to move within 2 years than renters who prefer to move within the rental sector. In the previous section it was calculated that among those who prefer to move to an owner-occupied home, homeowners are about as likely as renters are to realize their intention to move within 2 years. Previous research has frequently shown that homeowners state an intention to move less often (see for example Rossi 1955) and that homeowners move less often than renters (see for example Helderman et al. 2004). However, this study shows that homeowners do not have more difficulties to realize an intention to move than renters. A direct comparison between the stated intentions and actual behavior for filterers shows that if homeowners have made the decision to move, they do not necessarily encounter more obstacles in realizing their intention to move than renters.

10.4.2.3 The Influence of Other Individual and Housing Characteristics

In line with other studies, older people and non-western immigrants have a smaller probability of realizing their intention to move than younger people and natives (for example, see Kan 1999). The probability of realizing an intention to move decreases with increasing age, up to the age of 65. People who perceived their health to be good have a greater probability of realizing an intention to move than those who perceived their health to be bad or reasonably good.

The realization of stated intentions to move is also affected by housing characteristics. Filterers living in either a spacious home or a single family home are estimated to be less likely to realize their intention to move within 2 years time than those living in either a neutral crowding situation, or an apartment. Satisfaction with the current home also matters for the realization of intentions to move. Dissatisfied filterers are 1.2 times more likely to move than filterers who are satisfied with their current home. Thus, being dissatisfied with the current home gives an extra stimulus to realizing an intention to move within a short period of time. In contrast, satisfaction with the neighborhood does not have a significant effect on the probability of moving.

10.5 Conclusion and Discussion

In this chapter I have addressed the discrepancy between stated intentions to move and actual moving behavior. Using a unique longitudinal data-set in which survey data from a large-scale survey for the Netherlands were enriched with register data, it was found that the relationship between intentions and behavior is far from straightforward. Only one third of those intending to move within the next 2 years realized this intention within this period. Over the same period of time, six percent of those without an intention to move did change residence.

Similar to the longitudinal approach, the cross-sectional approach also shows a discrepancy between stated intentions to move and actual moves. However, a comparison between the longitudinal and the cross-sectional approach makes clear that this discrepancy is much larger on the individual level than on the aggregate level. Cross-sectional analyses are unable to show the underlying variation in the extent to which people behave according to their earlier stated intentions to move – most people do not realize their intention to move, while others move although they did not intend to do so – and this can produce incorrect results (compare Davies and Pickles 1985). When one wants to gain insight into the extent to which people behave according to their prior stated intentions to move, the longitudinal research method is therefore to be preferred over the cross-sectional research method.

In contrast with cross-sectional studies, longitudinal studies can also provide more insight into the extent to which various determinants influence the moving behavior of those with an intention to move. The longitudinal analysis presented in this chapter reveals that the probability of moving is particularly influenced by the urgency of the intention to move. In addition, housing preferences play a role; particularly those who want to move out of homeownership have a high probability of realizing their intention to move. Surprisingly, financial resources (related to income, education and employment situation) had hardly any effect on the probability of moving among intended filterers. The divergence between intentions and behavior can partly be ascribed to restrictions at the micro level (housing and household situation) and a lack of housing opportunities. It has been shown that people whose search location is in the densely populated Randstad area are 0.8 times as likely to translate intentions into action as those who prefer a home in the national periphery. This finding corresponds with the fact that the shortage of housing is especially problematic in the Randstad and stresses the relevance of contextual factors for the extent to which intentions to move will be realized. With the unique, large-scale longitudinal data-set used in this research it is actually possible to unravel the role of specific local housing market circumstances in the step from mobility intentions to actual mobility behavior. For this purpose, future research should include detailed information in the analysis regarding the local housing market conditions in the area where people intend to search for a home (for example, about the price per square meter, or about the average waiting time needed for a social rental home) using a multilevel design.

The results presented in this chapter contribute to our understanding of why people do not behave according to their initial stated intentions to move and give insight into the circumstances that hamper or stimulate the realization of stated intentions to move. However, I was not able to show to what extent the discrepancy between stated intentions to move and actual moving behavior can be ascribed to the fact that intentions might arise or fade due to constraints and facilitators previously ignored by the individual (Lu 1998) or to the occurrence of unforeseen life events. Previous research has indicated that unforeseen life events might also trigger moves among those who did not previously state an intention to change residence (for example see Kan 1999). Further research is needed to find out to what extent the occurrence of unforeseen life events after someone has reported whether he or she has an intention to move or not, may lead people to adjust their initial stated intention, resulting in a discrepancy between stated intentions and actual behavior.

Insight into the discrepancy between stated intentions to move and actual moving behavior can be very useful to policy makers since, in the Netherlands, intentions to move are used in housing demand estimations (see for example Den Otter 2007; VROM 2007d). It has been shown that the discrepancy between stated intentions and actual behavior is much smaller on the aggregated level than on the individual level; 'unexpected' stays are partly counterbalanced by 'unexpected' moves. This might be an indication that, on the aggregated level, intentions to move are an adequate indicator for predicting total moves (Van Hoorn 2009). However, stated intentions to move or to stay are certainly not perfect predictors of actual moving behavior. One might, therefore, argue that it is difficult to use the stated intentions of individuals for housing demand estimations; the more so since the characteristics of 'unexpected' movers do not always correspond with the characteristics of 'unexpected' stayers (Everaers and Lamain 1989).

Finally, it has been shown that a lack of housing opportunities and the fact that most filterers do not have an urgent need to move contributes to the discrepancy between stated intentions and actual behavior among filterers. If filterers cannot find a new home that meets their housing preferences, they might be more likely to postpone or cancel their intention to move than to accept a home that does not meet their initial housing preferences. Further investigations of the circumstances that hamper or stimulate the realization of stated intentions to move can be used to improve the effectiveness of housing assistance programs targeting particular population groups as well as to aid in the design of new programs (Lu 1999).

Acknowledgments I am very grateful to my two supervisors, Clara Mulder and Dorien Manting, for their highly valuable comments on earlier drafts of this chapter. Furthermore, I would like to thank Frank van Oort for his help with the macro-zoning of regions in the Netherlands, and Statistics Netherlands who allowed us to use data from the Social Statistical Database (SSD). The use of the SSD was possible due to the joint cooperation of Statistics Netherlands and the Netherlands Environmental Assessment Agency.

Appendix 1 Descriptive statistics of dependent and independent variables in the multivariate analysis for filterers (N = 11730)

	Frequency (%)	Of whom moved (%)
Moving behavior (dependent variable)		
Did not move within 2 year period	8,019 (68.4)	
Moved within 2 year period	3,711 (31.6)	
Characteristics of intention to move and housing preferences		
Urgency of intention to move		
Less urgent intention	9,281 (79.1)	25.7
Urgent intention	2,449 (20.9)	54.3
Reason for moving		
Demographic event	1,713 (14.6)	33.7
Housing	6,129 (52.3)	31.2
Work or education	859 (7.3)	33.1
Other reason	3,029 (25.8)	31.0
Preferred tenure		
Wish to rent	5,165 (44.0)	30.3
Wish to own	6,565 (56.0)	32.7
Preferred type of housing		
Wish for apartment	3,231 (27.5)	30.8
Wish for single family home	8,499 (72.5)	32.0
Preferred search location		
Randstad	5,485 (46.8)	30.8
Intermediate zone	3,393 (28.9)	31.7
National periphery	2,852 (24.3)	33.1
Resources and restrictions		
Income		
Lowest quartile	2,646 (22.6)	30.0
Second quartile	3,272 (27.9)	29.8
Third quartile	2,914 (24.8)	32.7
Highest quartile	2,898 (24.7)	34.1

(continued)

Appendix 1 (continued)

	Frequency (%)	Of whom moved (%)
Level of education		
Up to lower secondary	4,414 (37.6)	28.2
Higher secondary or medium vocational	3,741 (31.9)	33.0
Higher vocational or university	3,575 (30.5)	34.5
Employment status		
Employed	7,897 (67.3)	33.3
Not employed	3,833 (32.7)	28.3
Expected household composition		
Single	2,975 (25.4)	29.2
Couple without children	4,207 (35.9)	35.7
Family with children	4,548 (38.8)	29.5
Current tenure		
Renter	6,972 (59.4)	32.4
Owner	4,758 (40.6)	30.5
Other individual and housing characteristics		
Age		
<25	980 (8.4)	43.7
25–34	3,959 (33.8)	38.5
35–44	2,843 (24.2)	27.4
45–54	1,728 (14.7)	23.6
55–64	1,133 (9.7)	21.8
≥65	1,087 (9.3)	30.0
Ethnicity		
Native-born	9,210 (78.5)	32.2
Non-western immigrant	1,470 (12.5)	28.8
Western immigrant	1,050 (9.0)	30.2
Perceived health		
Healthy	9,607 (81.9)	32.8
Reasonably healthy	1,135 (9.7)	25.2
Less healthy	988 (8.4)	27.3
Type of home		
Apartment	5,341 (45.5)	35.6
Single family home	6,389 (54.5)	28.3
Crowding		
Crowded	2,425 (20.7)	34.4
Not crowded	3,286 (28.0)	32.8
Spacious	6,019 (51.3)	29.9
Satisfaction with home		
Satisfied	8,224 (70.1)	30.6
Not satisfied or dissatisfied	1,924 (16.4)	31.4
Dissatisfied	1,582 (13.5)	37.4
Satisfaction with neighborhood		
Satisfied	7,706 (65.7)	31.9
Neither satisfied nor dissatisfied	1,715 (14.6)	29.6
Dissatisfied	2,309 (19.7)	32.4

Source: SSD Satellite Spatial and Social Mobility 1999–2005, including HDS 2002

References

Ajzen, I. (1985). From intentions to actions: A theory of planned behavior. In J. Kuhl & J. Beckmann (Eds.), *Action control. From cognition to behavior* (pp. 11–35). Berlin: Springer.

Ajzen, I. (1991). The theory of planned behavior. *Organizational Behavior and Human Decision Processes, 50*, 179–211.

Ajzen, I., & Madden, T. J. (1986). Prediction of goal-directed behavior: Attitudes, intentions, and perceived behavioral control. *Journal of Experimental Social Psychology, 22*, 453–474.

Allison, P. D. (1999). *Logistic regression using the SAS system: Theory and application*. Cary: SAS Institute.

Bakker, B. F. M. (2002). Statistics Netherlands' approach to social statistics: The social statistical dataset. *OECD Statistics Newsletter, 11*, 4–6.

Behr, A., Bellgardt, E., & Rendtel, U. (2005). Extent and determinants of panel attrition in the European community household panel. *European Sociological Review, 21*(5), 489–512.

Brown, L. A., & Moore, E. G. (1970). The intra-urban migration process: A perspective. *Geografiska Annalar, 52*(1), 1–13.

Buck, N., Gershuny, J., Rose, D., & Scott, J. (Eds.). (1994). *Changing households: The BHPS 1990 to 1992, ESRC research centre on micro-social change*. Colchester: University of Essex.

Clark, W. A. V., & Dieleman, F. M. (1996). *Households and housing: Choice and outcomes in the housing market*. New Brunswick: Center for Urban Policy Research.

Clark, W. A. V., & Onaka, J. L. (1985). An empirical test of a joint model of residential mobility and housing choice. *Environment and Planning A, 17*, 915–930.

Crowder, K. D. (2001). Racial stratification in the actuation of mobility expectations: Microlevel impacts of racially restrictive housing markets. *Social Forces, 79*(4), 1377–1396.

Davies, R. B., & Pickles, A. R. (1985). Longitudinal versus cross-sectional methods for behavioural research: A first-round knockout. *Environment and Planning A, 17*(10), 1315–1329.

De Groot, C., Manting, D., & Boschman, S. (2008). *Verhuiswensen en verhuisgedrag in Nederland. Een landsdekkend onderzoek*. The Hague: PBL Netherlands Environmental Assessment Agency.

De Jong, G. F. (1994). *Choice process in migration intentions and behavior, working paper 95-04*. Pennsylvania: Pennsylvania State University, Population Research Institute.

Den Otter, H. J. (2007). *Een nieuwe benadering van het woningtekort [A new approach of the housing demand estimation]*. Delft: ABF Research.

Dieleman, F. M. (1996). Modeling housing choice. *Netherlands Journal of Housing and the Built Environment, 11*(3), 201–207.

Duncan, G. J., & Newman, S. J. (1976). Expected and actual residential mobility. *Journal of the American Institute of Planners, 42*, 174–186.

Everaers, P., & Lamain, C. (1989). Gewenst en gerealiseerd verhuisgedag, een analyse met behulp van het Woningbehoeftenpanel 1988 [expected and actual residential mobility, an application of results of the housing demand panel 1988]. *Supplement bij de Sociaal-Economische Maandstatistieken, 2*, 16–20.

Feijten, P. (2005). Union dissolution, unemployment and moving out of homeownership. *European Sociological Review, 21*(1), 59–71.

Feskens, R., Hox, J., Lensvelt-Mulders, G., & Schmeets, H. (2006). Collecting data among ethnic minorities in an international perspective. *Field Methods, 18*(3), 284–304.

Fishbein, M., & Ajzen, I. (1975). *Belief, attitude, intention, and behavior: An introduction to theory and research*. Reading: Addison-Wesley Publishing Company.

Garson, G.D. (2009). *Logistic regression*. Statnotes: Topics in multivariate analysis, http://faculty.chass.ncsu.edu/garson/PA765/logistic.htm. Retrieved 22 April 2009.

GfK. (2009). *Ruimte op de woningmarkt: Onderzoek naar de realisatie van verhuiswensen*. The Hague: Ministerie van VROM.

Goetgeluk, R. W. (1997). *Bomen over wonen: Woningmarktonderzoek met beslissingsbomen [Trading off housing preferences: Housing market research with decision plan nets]*. Utrecht: Faculteit Ruimtelijke Wetenschappen Universiteit Utrecht.

Goetgeluk, R. W., & Hooimeijer, P. (1991). *Household formation and access to housing in the Netherlands*. Paper prepared for the conference "Housing Policy as a Strategy for Change", Oslo, 24–27 June 1991.

Goetgeluk, R. & Hooimeijer, P. (2002). *The evaluation of Decision Plan Nets for bridging the gap between the ideal dwelling and the accepted dwelling*. CYBERGEO, 226, www.cybergeo.eu.

Goetgeluk, R. W., Hooimeijer, P. & Dieleman, F. M. (1992). *The effectiveness of housing search; The role of motives for moving and housing market adjustment*. Paper prepared for the conference "European Cities: Growth & Decline", Den Haag, 13–16 April 1992.

Gouweleeuw, J. (2006). Hoe toegankelijk is het Sociaal Statistisch Bestand (SSB). In F. Bastiaans, L. Engberts, & F. Linder (Eds.), *Sociale samenhang in beeld, het SSB nu en straks* (pp. 189–201). Voorburg: Statistics Netherlands.

Gray, A. (2001). *Definitions of crowding and the effects of crowding on health: A literature review*. Wellington: Ministry of Social Policy.

Haffner, M., Boumeester, H., Dol, K., Goetgeluk, R., & Neuteboom, P. (2008). *Woonuitgaven 2002–2006 in beeld*. Delft: OTB, Technical University of Delft.

Helderman, A. C., Mulder, C. H., & van Ham, M. (2004). The changing effect of home ownership on residential mobility in the Netherlands, 1980–1998. *Housing Studies, 19*(4), 601–616.

Hooimeijer, P., & Linde, M. A. J. (1988). *Vergrijzing, individualisering en de woningmarkt; Het WODYN-simulatiemodel*. Utrecht: Geografisch Instituut Rijksuniversiteit Utrecht.

Hooimeijer, P., & Oskamp, A. (1996). A simulation model of residential mobility and housing choice. *Journal of Housing and the Built Environment, 11*(3), 313–336.

Hooimeijer, P., & Poulus, C. (1995). *Rapportage woningbehoefteonderzoek 1993/1994*. The Hague: Ministry of Housing, Spatial Planning and the Environment.

Hosmer, D. W., & Lemeshow, S. (1989). *Applied logistic regression*. New York: Wiley.

Jaccard, J. (2001). *Interaction effects in logistic regression* (Sage University papers series on quantitative applications in the social sciences, 07-135). Thousand Oaks: Sage.

Kan, K. (1999). Expected and unexpected residential mobility. *Journal of Urban Economics, 45*, 72–96.

Kearns, A., & Parkes, A. (2003). Living and leaving poor neighbourhood conditions in England. *Housing Studies, 18*(6), 827–851.

Kim, J. H., Pagliara, F., & Preston, J. (2005). The intention to move and residential location choice behaviour. *Urban Studies, 42*(9), 1621–1636.

Kleinhans, R. J. (2005). *Sociale implicaties van herstructurering en herhuisvesting [Social implications of urban restructuring and relocation]*. Delft: Delft University Press.

Konter, W., & van den Booren, H. (1988). *Hoe zwaar mag een verhuiswens wegen? Een onderzoek naar de realisatie van verhuis- en woonwensen van niet-gehuwde jongeren tot 35 jaar en ouderen van 55 jaar en ouder*. Haarlem: Werkgroep Regionaal en Lokaal Woningmarktonderzoek in Noord-Holland.

Kruythoff, H. (1994). Starters in the housing market in an urban region: The case of the Randstad Holland, a diversified housing-shortage area. *Housing Studies, 9*(2), 219–244.

Landale, N. S., & Guest, A. M. (1985). Constraints, satisfaction and residential mobility: Speare's model reconsidered. *Demography, 22*(2), 199–222.

Lu, M. (1998). Analyzing migration decisionmaking: Relationships between residential satisfaction, mobility intentions, and moving behavior. *Environment and Planning A, 30*, 1473–1495.

Lu, M. (1999). Do people move when they say they will? Inconsistencies in individual migration behavior. *Population and Environment, 20*(5), 467–488.

Madden, T. J., Ellen, P. S., & Ajzen, I. (1992). A comparison of the theory of planned behavior and the theory of reasoned action. *Personality and Social Psychology Bulletin, 18*(1), 3–9.

Meeuwissen, B., Ferment, B., van Huijsduijnen, J. H., van Til, R. J., & Hooft, J. (2003). *WoningBehoefte Onderzoek 2002* (Onderzoeksdocumentatie [Housing demand survey 2002. Research documentation]). Delft: ABF Research.

Moore, E. G. (1986). Mobility intention and subsequent relocation. *Urban Geography, 7*(6), 497–514.
Mulder, C. H. (1993). *Migration dynamics: A life course approach*. Amsterdam: Thesis Publishers.
Mulder, C. H. (1996). Housing choice. Assumptions and approaches. *Journal of Housing and the Built Environment, 11*(3), 209–232.
Mulder, C. H., & Hooimeijer, P. (1999). Residential relocations in the life course. In L. van Wissen & P. Dykstra (Eds.), *Population issues: An interdisciplinary focus* (pp. 159–186). New York: Plenum Press.
Murie, A. (1974). *Household movement and housing choice* (Vol. 28). Birmingham: Centre for Urban and Regional Studies, University of Birmingham.
Myers, D., Baer, W. C., & Choi, S. Y. (1996). The changing problem of overcrowded housing. *Journal of the American Planning Association, 62*(1), 66–84.
Oskamp, A. (1997). *Local housing market simulation: A micro approach*. Amsterdam: Thela Thesis.
Parkes, A., & Kearns, A. (2003). Residential perceptions and housing mobility in Scotland: An analysis of the longitudinal scottish house condition survey 1991–96. *Housing Studies, 18*(5), 673–701.
Priemus, H. (1984). *Verhuistheorieën en de verdeling van de woningvoorraad*. Delft: Delft University Press.
REA (Council of Economic Advisers) (2006). De woningmarkt uit het slot, Tweede Kamer vergaderjaar 2005–2006, 30 507 nr. 2. The Hague: Sdu Uitgevers.
Rossi, P. H. (1955). *Why families move. A study in the social psychology of urban residential mobility*. Glencoe: Free Press.
Ruspini, E. (1999). Longitudinal research and the analysis of social change. *Quality and Quantity, 33*, 219–227.
Sheeran, P. (2002). Intention-behavior relations: A conceptual and empirical review. *European Review of Social Psychology, 12*(1), 1–36.
Speare, A. (1974). Residential satisfaction as an intervening variable in residential mobility. *Demography, 11*(2), 173–188.
Statistics Netherlands. (2006). *Statistics Netherlands*. Voorburg: Statistics Netherlands.
Timmermans, H., Molin, E., & van Noortwijk, L. (1994). Housing choice processes: Stated versus revealed modelling approaches. *Netherlands Journal of Housing and the Built Environment, 9*(3), 215–227.
Van Groenigen, C., & van der Veer, J. (2006). *Slaagkansenmonitor 2005*. Amsterdam: Gemeente Amsterdam, Dienst Wonen/Amsterdamse Federatie van Woningcorporaties.
Van Hoorn, F. (2009). Doen mensen ècht niet wat zij zeggen? *Tijdschrift voor de Volkshuisvesting, 3*, 41–46.
Van Kempen, R., Dekker, F., & Floor, J. (1990). The desire to move and residential mobility. In J. van Weesep & P. Korcelli (Eds.), *Residential mobility and social change* (Studies from Poland and the Netherlands, pp. 84–94). Amsterdam/Utrecht: KNAG/Geografisch Instituut Rijksuniversiteit Utrecht).
Van Oort, F. (2003). *Urban growth and innovation* (Analysis of spatially bounded externalities in the Netherlands). Aldershot: Ashgate.
VROM (Ministry of Housing, Spatial Planning and the Environment) (2007a). *Bouwen voor de schuifpuzzel*. The Hague: Ministerie van VROM.
VROM (2007b). *Wonen op een rijtje* (De resultaten van het WoonOnderzoek Nederland 2006). The Hague: Ministerie van VROM.
VROM (2007c). *Woningmarktverkenningen* (Socrates 2006). The Hague: Ministerie van VROM.
VROM (2007d). *Primos prognose 2007* (De toekomstige ontwikkeling van bevolking, huishoudens en woningbehoefte). The Hague: Ministerie van VROM.
VROM (2003). *Gescheiden markten? De ontwikkeling op de huur- en koopwoningmarkt*. The Hague: Ministerie van VROM.
Winship, C., & Radbill, L. (1994). Sampling weights and regression analysis. *Sociological Methods and Research, 23*, 230–257.
Yamaguchi, K. (1991). *Event history analysis*. Newbury Park: Sage Publications.

Chapter 11
Discussion and Directions for Future Research

Sylvia J.T. Jansen, Henny C.C.H. Coolen, and Roland W. Goetgeluk

11.1 Introduction

This book contains a description of nine methods and analytical techniques that are currently applied in housing preference research. In order to help professionals to select the most appropriate method or to judge earlier studies on their merits, we introduced three dimensions that concerned: 1: the origin of the data (stated or revealed), 2: the freedom of attribute choice, and, 3: attribute-based versus alternative-based approach. These dimensions were selected because they reflect broad differences between the nine methods and techniques. The dimensions were described in the Introduction. In the current chapter, potential limitations related to these three dimensions are discussed. Note that the list of potential limitations is not necessarily complete and exhaustive. Furthermore, limitations that apply to individual methods are not mentioned as they are discussed in their corresponding chapters. Also, potential benefits of the three dimensions are not discussed in detail; a limitation of one aspect of the particular dimension (for example, methods with freedom of attribute choice may be costly to administer) can be the benefit of another method (for example, methods with no freedom of attribute choice may be inexpensive to administer). Providing both limitations and benefits at the same time would therefore provide extensive overlapping. Finally, as we also mentioned in the Introduction chapter, what consumers want can be measured in many different ways. Which particular method has to be chosen can only be answered in the light of the purpose of the measurement (Hooimeijer 1994). Different methods lead to different outcomes. The choice for a specific method cannot therefore be based on

S.J.T. Jansen (✉) and H.C.C.H. Coolen
OTB Research Institute for the Built Environment,
Delft University of Technology, Delft, The Netherlands
e-mail: s.j.t.jansen@tudelft.nl; h.c.c.h.coolen@tudelft.nl

R.W. Goetgeluk
Demography & Housing, ABF Research, Delft, The Netherlands
e-mail: roland.goetgeluk@abf.nl

S.J.T. Jansen et al. (eds.), *The Measurement and Analysis of Housing Preference and Choice*, DOI 10.1007/978-90-481-8894-9_11,

Table 11.1 Overview of methods and analytical techniques with regard to the three dimensions

Applies to:	Origin	Design	
Dimensions	Stated or revealed	Freedom of attribute choice	Compositional versus decompositional
Traditional housing Demand research method	Stated	No	Compositional
Decision plan nets method	Stated	Yes	Compositional
Meaning structure method	Stated	Yes	Compositional
Multi-attribute utility method	Stated	Yes	Compositional
Conjoint analysis method	Stated	No	Decompositional
Residential images method	Stated	No	Decompositional
Lifestyle method	Stated	No	N.a.
Neo-classical economic analysis	Both	No	N.a.
Longitudinal analysis	Both	No	N.a.

N.a. not applicable

the methodological superiority of one method over another but should be directed by the type of information in which one is interested (Hooimeijer 1994). Our discussion of potential limitations is therefore only meant as an, albeit imperfect, guideline to help professionals to choose the appropriate method or technique for the specific situation or to decide upon the justification of conclusions of studies that have been performed in practice.

11.2 Comparison of Methods and Analytical Techniques with Regard to the Three Dimensions

Table 11.1 shows the description of the methods and techniques according to the three dimensions that were distinguished in the Introduction. In the following section we will discuss the potential limitations of the three dimensions. We start with the dimension of stated or revealed preferences.

11.2.1 Dimension 1: Stated or Revealed Preferences

The first dimension that was introduced concerned the origin of the data, namely does it concern choices that have actually been made in the 'real world' (revealed preferences) or stated choices and preferences in response to survey questions (stated preferences)? A potential limitation of the first approach is that it assumes that revealed preferences reflect underlying preferences (Timmermans et al. 1994).

However, outcomes in the housing system may frequently reflect the dominance of constraints, such as income and imperfect information, rather than preferences (Maclennan 1977). For example, consumers may choose to live in a multi-family dwelling because of budget or market constraints (availability) and not because they really want to. The actual housing situation is always an interaction of constraints and preferences, especially in the lower value ranges of the housing system, and it is very difficult to disentangle preferences from restraints.

Aside from this limitation, the revealed preference approach assumes that consumers always make rational choices and seek optimum solutions (Maclennan 1977). Thus, it is assumed that a consumer makes an explicit and rational choice for a particular dwelling out of all the available options. But in reality the consumer may not be aware of all the available options or may choose a particular dwelling on less rational grounds, so this assumption may not hold in practice. Furthermore, some explanatory factors that are important but of which the researcher is unaware or may be unable to include might be omitted from the analyses on revealed preferences. The analyses may also not capture effectively the impact of uncommon attributes or unusual attribute levels (Earnhart 2002). Another limitation of the revealed preference approach concerns the finding that the attributes of housing alternatives in real markets may show high correlations, for example, bigger houses are typically more expensive. And high correlations between predictors may lead to misleading estimates as a result of (near) multicollinearity (Molin et al. 1996). Finally, alternatives that do not currently exist in the real world or attribute levels that go beyond the range of current experience, cannot be analyzed using revealed preferences (Walker et al. 2002).

In contrast, it has been argued that the stated preference approach might not be valid because people can express temporary wants or ideals that cannot be realized in the actual housing market (Vriens 1997; Heijs et al. 2009). This criticism only makes sense, though, if one does not make a distinction between preferences and choices. If, on the other hand, one distinguishes preference, as the relative attractiveness of a feature, from choice, as actual behavior, then stated preferences may not have a strong relationship with actual housing market behavior. Furthermore, the stated preference approach assumes that respondents are able to articulate their preferences whereas they can be indifferent or their preferences may depend on particular conditions (Molin et al. 1996). Finally, the validity of the responses is a concern as stated preferences may be influenced by factors such as social desirability, risky decision-making and cognitive dissonance reduction (Molin et al. 1996; Walker et al. 2002).

A summary of the above mentioned potential limitations of both approaches is provided in Table 11.2.

11.2.2 Dimension 2: Freedom of Attribute Choice

The second dimension concerns freedom of attribute choice. A method that allows freedom of attribute choice can be applied (but need not necessarily so) in such a

Table 11.2 Summary of the potential limitations of stated and revealed preference methods

Revealed preference approach

- It may be difficult to disentangle real preferences from (market) constraints.
- The method assumes that consumers always make rational choices and seek optimum solutions, which is questionable.
- Important explanatory factors may be omitted from the analysis.
- Attributes may show high correlations, which can lead to misleading coefficients.
- Alternatives that do not currently exist in the real world or attribute levels that go beyond the range of current experience, cannot be analyzed.

Stated preference approach

- The method might not be valid because people can express temporary wants or ideals that cannot be realized in the actual housing market.
- The method assumes that respondents are able to articulate their preferences whereas they can be indifferent or their preferences may depend on particular conditions.
- The method may be influenced by factors such as social desirability, risky decision-making and cognitive dissonance reduction.

Table 11.3 Summary of potential limitations of methods that provide freedom of attribute choice and those that do not

Freedom of attribute choice

- The data is usually collected with face-to-face interviews or telephone interviews, which is relatively time-consuming and costly.
- Results may become rather idiosyncratic, which means that it can be difficult to report on general preferences.

No freedom of attribute choice

- Important attributes that were not included in the research design may be lacking.

way that respondents can choose their own salient attributes. The potential limitation of a method using freedom of attribute choice is that the data is usually collected using face-to-face interviews or telephone interviews, which are relatively time-consuming and costly. Furthermore, the results obtained using such methods may become rather idiosyncratic, as every respondent can contribute his/her individual dwelling (environment) attributes to the decision-making process. This means that it can be difficult to report on general preferences. A potential limitation of methods that do not provide freedom of attribute choice is that they use pre-coded questions based on a limited number of selected attributes of the dwelling (environment). Important attributes that were not included in the research design may be lacking. The potential limitations mentioned are summarized in Table 11.3.

11.2.3 Dimension 3: Compositional Versus Decompositional Approach

The third dimension relates to whether the measurement method is attribute-based (compositional) or alternative-based (decompositional). With a decompositional approach a dwelling profile is evaluated as a whole. Parameters are derived statistically from the decision-maker's holistic evaluative responses to profile descriptions designed by the researcher. For the compositional method, housing preferences are explored by recording separately and explicitly how people evaluate housing attributes. The importance of each attribute can be weighted and can be combined with the values, using an assumed algebraic rule, to arrive at an overall evaluation.

The strength of the compositional method is its simplicity: the measurement task is relatively easy and straightforward. A potential limitation of this method is that the researcher must specify a priori the mathematical function that will be used to combine the separate evaluations (Veldhuisen and Timmermans 1984). Usually, the simple additive combination rule is applied, which is explained in the chapter on the Multi-Attribute Utility method. It implies that a value for a separate attribute level, for example, a semi-detached dwelling, is multiplied with the importance of that particular attribute, for example dwelling type, and that these weighted attribute values are summed over all attributes of the particular dwelling. Besides this particular combination rule, others are possible, such as the multiplicative and the multi-linear combination rule. A weakness of the compositional approach is that the appropriateness of the selected combination rule can only be tested if data for some external criterion is available. In contrast, with the decompositional approach the form of the combination rule can be explicitly tested by comparing the derived solution to the observed data (the overall preference for a particular dwelling profile). The form of the model can be tested statistically because it has an associated error theory (the difference between the estimated and actually observed overall preference).

A second potential limitation of the compositional method is that a trade-off between attributes need not be involved. The approach implicitly assumes that respondents can express their evaluation of a distinct dwelling (environment) attribute irrespective of the levels of other attributes (Timmermans et al. 1994; Molin et al. 1996). For example, it assumes that the attribute of price can be valued without knowing the size of the dwelling. This assumption is questionable in the case of choice, but does not need to be a problem in the case of preference, since the preferred dwelling may consist of the collection of preferred attributes and their accompanying levels. However, some argue that it is in making trade-offs between more of one thing and less of another that one's values are most often revealed to oneself and to outside observers (Payne et al. 1999). For example, a large living room is desirable, but is it still desirable if it comes with increased financial costs? In this sense, making trade-offs is a crucial aspect of high-quality, rational decision-making (Payne et al. 1999).

However, in practice, decision makers may often avoid making explicit trade-offs because this is cognitively and emotionally burdening (Payne et al. 1999). This is a potential limitation of the decompositional methods. People cannot simultaneously integrate a great deal of information. Respondents may adopt fairly simple procedures and rules (heuristic strategies), which reduce cognitive overload. For example, respondents may select the status quo option or only pay attention to the most important attributes while ignoring the rest (Lindberg et al. 1989; Gregory et al. 1993; Chang and Liu 2008). Selecting the status quo option may, for example, be reflected in a choice not to move in the case of a hypothetical choice in a conjoint measurement task between "Move to dwelling A", "Move to dwelling B" and "Stay in the current dwelling". Only paying attention to, for example, the price of a dwelling, when several other attribute levels are also presented, is an example of attending only to the most important attributes. The choice task is simplified by the respondent by ignoring the information provided in the other attribute levels.

Vriens (1997) sums up a number of additional potential limitations of the classical compositional approach (which resembles most the Multi-Attribute Utility method): (1) Research has shown that the importance of important attributes is underweighted by respondents whereas the importance of relatively unimportant attributes is overweighed; (2) The direct measurement of importance ratings might elicit socially desirable responses; (3) Interaction effects cannot be measured.

Vriens (1997) also mentions a number of potential limitations of the decompositional approach (specifically: conjoint measurement): (1) the measurement task cannot easily be done by telephone interview; (2) special procedures have to be performed when the number of attributes or attribute levels becomes too high; and (3) performing such a study is generally more costly both in terms of time and money than studies using a compositional approach. Walker et al. (2002) explain that errors might arise when an inefficient or inappropriate design is used with the decompositional method. If the variations in attribute levels offered are too small or too large or if unrealistic attribute levels or unrealistic combinations of attribute levels are presented, respondents may provide suboptimal responses.

Note that in the current overview the Multi-Attribute Utility method and the Decision Plan Net method are considered to be compositional methods, because the separate attributes are the starting point in the measurement procedure. However, in the Multi-Attribute Utility method, techniques that are based on trade-offs can be applied, such as swing weights. Here, the importance of an attribute is determined by comparing dwelling profiles that "swing" between the worst and the best level of a particular attribute. The extent to which the swings in each attribute contribute to overall value differences is examined (Payne et al. 1999). This way, trade-offs between attributes can be determined. Similarly, in the last step of the Decision Plan Nets method, the respondent is asked to rank appropriate alternatives according to preference. Implicitly, a trade-off between attributes is made. So, the difference between methods that involve trade-offs and those that do not, is not as clear-cut as it might seem at first sight. See Table 11.4 for a summary.

11.2.4 Compensatory Versus Non-Compensatory Methods

As a fourth, but not distinguishing dimension, we mentioned in the Introduction chapter the difference between compensatory and non-compensatory methods. Compensatory decision-making implies that a low value on one attribute can be compensated by a high value on one or more other attributes. Thus, the specific

Table 11.4 Summary of potential limitations of compositional and decompositional methods

Compositional approach

- The appropriateness of the mathematical function that is used to combine the separate evaluations cannot be tested unless data on some external criterion is available.
- A tradeoff between attributes need not be involved. This may question the validity of the method.
- The importance of important attributes may be underweighted whereas the importance of relatively unimportant attributes may be overweighed.
- The direct measurement of importance ratings might elicit socially desirable responses.
- Interaction effects cannot be measured.

Decompositional approach

- Decision makers may avoid making explicit tradeoffs and may adopt simplifying heuristics because the measurement task is cognitively and emotionally burdening.
- The measurement task cannot easily be done by telephone interview; performing such a study is generally more costly both in terms of time and money.
- Special procedures have to be performed when the number of attributes or attribute levels becomes too high.
- Variations in attribute levels that are too small or too large, unrealistic attribute levels or unrealistic combinations of attribute levels may lead to suboptimal responses.

alternative may still obtain a high overall evaluation score despite a low value on one or more attributes. In contrast, a non-compensatory decision method implies that a highly valued attribute cannot make up for a weak valued one. The valuation of an attribute above or below a certain preferred threshold therefore must lead to the rejection of an alternative vacancy.

Our reason for not including this dimension in Table 11.3 in the Introduction is that we believe that almost all methods can be used in a compensatory or non-compensatory way, depending on how the questions are framed or on how the analysis is performed. For example, in the Multi-Attribute Utility method a linear additive function can be used to describe compensatory decision strategies. This means that evaluations for separate attribute levels are simply added to obtain an overall utility for a particular dwelling. A low evaluation for a particular attribute level can be compensated by high evaluations for other attributes. However, a multiplicative function, which may approximate non-compensatory preference structures, can also be applied. This means that low evaluations can hardly be compensated for.

Furthermore, for the less statistically sophisticated methods, whether or not some method is compensatory might depend on whether the trade-off of preferences is questioned. If respondents are allowed to reject an alternative on the basis of its level of functioning on one or more attributes, the method is used in a non-compensatory way. If they were not allowed to reject alternatives, the method is used in a compensatory way.

Lindberg et al. (1989) argue that compensatory methods are not tenable if respondents use simplifying heuristics to make decisions. One example of such a simplifying heuristic is the lexicographic decision rule. It implies that the decision-maker determines the most important attribute and then examines all the alternatives for that attribute. The alternative with the best value on the most important attribute is chosen. For example, attention is given only to the attribute of price and all dwelling alternatives are selected on the basis of this attribute only. In a recent study Dieckmann et al. (2009) showed that whether a compensatory or non-compensatory (in this case: lexicographic rule) decision rule is used may be dependent upon the mode of measurement. The authors found that a lexicographic model was better in predicting ranking data whereas a basic weighted additive model was better with rating data. They attributed this result to the greater complexity of the ranking task (comparing 18 alternatives simultaneously) than of the rating task (providing a rating for each alternative at the time). Furthermore, a weighted additive model performed better when only a small number of alternatives or attributes was involved. A more complex task may increase the need for simplifying heuristics, such as the lexicographic method. The authors concluded that a relatively large number of alternatives or attributes may induce a shift from compensatory to non-compensatory processing in order to reduce the number of relevant alternatives as quickly as possible (Dieckmann et al. 2009). However, decision-makers may also use both decision strategies. They may begin by applying a non-compensatory decision rule to eliminate alternatives that do not meet the criteria for the most important attributes, such as price and location. Next, they may use a compensatory decision rule to evaluate the remaining alternatives across a wide range of less important criteria.

11.3 Directions for Future Research

Our overview of methods and analytical techniques currently used in housing preference research leads us to conclude that considerable advances have been made on the issues related to the methodology of measuring housing preference and choice. However, there is still much work that has to be done. A number of those topics are outlined below.

11.3.1 The Process of Problem-Solving

In the stated preference method it is assumed, albeit implicitly, that consumers have articulated values, goals and plans, which means that they know their preferences directly. As a consequence the different approaches are mainly concerned with the best method of eliciting these preferences. Because of the assumption of articulated preferences the focus in all current approaches to housing choice and housing preference is almost entirely on the act of decision making (what is chosen or what is preferred), while hardly any attention is given to the process of problem solving (how decisions are reached). This remains an important topic for future research. More information on the construction of preferences can, for example, be found in Lichtenstein and Slovic (2006).

11.3.2 The Measurement Instrument

The measurement instrument has undergone some changes over time. Whereas in the past written questionnaires, telephone interviews and face-to-face interviews were frequently employed, there is now a fast-growing trend towards using web-based questionnaires. The latter instrument has the benefits of yielding appropriate data, decreases the cognitive burden for the respondent, enables images to be included and, possibly most important of all, it is relatively cheap.

By yielding appropriate data we mean that requirements can be set in order to obtain correct answers. For example, when exactly two options out of a number of options have to be chosen, this requirement can be imposed upon the response. Of course, such requirements can also be enforced using telephone or face-to-face interviewing, but it is more difficult in the case of a written questionnaire. The cognitive burden of filling out a questionnaire can be decreased by the option to build in routing questions in a web-based questionnaire. This way it can be ensured that respondents only answer questions that apply to them. Besides, the fact that a web-based questionnaire can be filled out at any desired point in time diminishes the cognitive burden. Respondents are not dependent upon planned interviews at prearranged points in time. The benefits with regard to cost are obtained from the fact that there are no interviewers needed to obtain the data. Furthermore, it does not require the data to be entered into a dataset as is the case with written questionnaires.

Besides the afore mentioned advantages, a web-based instrument is appropriate for administering relatively difficult questions, because of the technical possibilities (as, for example, including routing and images). An example is the study by Boumeester et al. (2008) in which relatively difficulty measurement methods, such as the Decision Plan Nets method and the Meaning Structure method, were administered using a web-based questionnaire. In this study also a special web-based instrument was developed to administer questions based on the Conjoint Analysis method, with the option of including images for each attribute (Picture Enabled Preference Survey Instrument (PEPSI): Boumeester et al. 2008).

Of course, there are also limitations to the use of web-based instruments. The most important of all is the selectivity of the respondent group. It is well-known that older people as well as non-western immigrants fill out web-based questionnaires less frequently than younger and native respondents. It is important that more research is carried out into solutions to increase the representativeness of the responses obtained with the use of web-based questionnaires.

11.3.3 Measuring Individual Preferences Versus Group Preferences

Housing preferences are usually elicited from individuals. These individuals, however, are frequently only part of a household. It is therefore questionable whether the preferences of individual respondents represent the opinions of the entire household that they are part of. Even if respondents are asked to consider the preferences of all the persons in the household in their response, it seems unlikely that they are aware of those preferences. And, if they are aware, if they are able and willing to provide weighted responses (Musterd 1989). The common practice to select one family member as a representative in housing surveys to provide responses that are supposed to reflect family judgment is unlikely to result in valid and reliable measurements of residential preferences (Molin et al. 1999). Despite profound work into measuring group-based preferences by, for example, Molin et al. (1997, 1999, 2002), we believe that this topic has still had too little attention paid to. We therefore encourage researchers to take account of the fact that preferences of different household members may differ from household preferences.

11.3.4 The Trend Towards Locally-Oriented Housing Preference Research

An important goal in housing management and policy is to improve the correspondence between housing demand and supply. The housing market in some regions has become more relaxed in the last decades. Some regions in the Netherlands even have to cope with a decreasing number of inhabitants, a higher level of residential turnover and increasing vacancies. Therefore, suppliers in these regions have to attune more to the needs and desires of (potential) residents in order to rent or sell their dwellings. This has led to a shift in direction from a supplier-market to a market that is more focused on demand. However, other regions are still experiencing house price growth and excess demand. This growth especially takes place in large cities and their surrounding villages. New building locations are scarce in urban areas and many dwellings will have to fit into the existing built environment. This asks for a building strategy that is aimed at building in higher densities, transforming residential environments and reuse of existing buildings and sites (Vromraad 2009). The simultaneous occurrence of both developments results in increasing regional differences in the housing market. This trend can also be seen in other countries, for example, in northern England (Nevin et al. 2001).

Besides, demographic, socio-economic and socio-cultural shifts have taken place in western economies in recent decades: households have become smaller and the variation in household types has increased. Other changes concern a greater variety of specific lifestyle-based subcultures and the expansion of the proportion of affluent households. These shifts have generated a broader variety in housing behavior (Kersloot and Kauko 2004).

As a consequence, study results can be generalized less easily and more research into housing preference and choice in specific locations is necessary. It implies that the way in which housing research is performed has to change from a focus on market constraints and population preferences to a focus on market possibilities and micro-level preferences. Thus, more attention is needed for locally-oriented housing preference research.

There are a number of ways to perform more locally-oriented research: (1) using a more qualitative (less-structured) approach, (2) including more geographical variation into the quantitative approach, and (3) using a lifestyle approach.

Kersloot and Kauko (2004) expect that there will be a rising demand for disaggregated and qualitative research tools because these tools are able to cope with a growing diversity of housing preferences. With qualitative tools they refer to methods to obtain less-structured data, such as casual observing, in-depth interviews and focus group discussions, and laddering (see chapter 4 in this volume).

Including more geographical variation into the quantitative approach can be performed in two ways. The first approach is to use large datasets, such as on house-prices and transactions (moves), and to include additional geographical details with a GIS approach, or otherwise. Applying multilevel models to such enriched datasets could deepen our insight into contextual effects. The second approach is to perform a meta-analysis on a number of smaller sets of stated preferences across localities with the purpose of revealing both more general trends and those that turn out to be more local.

As a third possibility to obtain locally oriented housing preferences, lifestyle research is put forward. For more information on lifestyle research, see Chap. 8.

11.4 Concluding Remarks

The method is a key factor in research. It is the link between the theory, goal and problem definition of research on the one hand and the results on the other hand. A mutual understanding of the basics of various methods and techniques is therefore a necessary condition to support research. The aim of this book is to provide the ins and outs of nine methods and analytical techniques commonly used in housing preference research. We introduced three important dimensions with which the various methods and analytical techniques can be compared. Together with the description of the goal and the type of outcome of each particular method, this information may be helpful in selecting the correct method to answer a particular research question or to decide upon the justification of the results of previously published studies into housing preferences. We hope that this book will be useful in fulfilling this purpose.

References

Boumeester, H. J. F. M., Coolen, H. C. C. H., Dol, C. P., Goetgeluk, R. W., Jansen, S. J. T., Mariën, A. A. A., & Molin, E. (2008). *Module Consumentengedrag WoON 2006, Hoofdrapport*. Delft: Onderzoeksinstituut OTB.

Chang, C. C., & Liu, H. H. (2008). Information format-option characteristics compatability and the compromise effect. *Psychology and Marketing, 25*(9), 881–900.

Dieckmann, A., Dippold, K., & Dietrich, H. (2009). Compensatory versus noncompensatory models for predicting consumer preferences. *Judgment and Decision Making, 4*(3), 200–213.

Earnhart, D. (2002). Combining revealed and stated data to examine housing decisions using discrete choice analysis. *Journal of Urban Economics, 51*(1), 143–169.

Gregory, R., Lichtenstein, S., & Slovic, P. (1993). Valuing environmental resources: A constructive approach. *Journal of Risk and Uncertainty, 7*, 177–197.

Heijs, W., Carton, M., Smeets, J., & van Gemert, A. (2009). The labyrinth of life-styles. *Journal of Housing and the Built Environment, 24*(3), 347–356.

Hooimeijer, P. (1994). Hoe meet je woonwensen? Methodologische haken en ogen. In I. Smid & H. Priemus (Eds.), *Bewonerspreferenties: Richtsnoer voor investeringen in nieuwbouw en de woningvoorraad* (pp. 3–12). Delft: Delftse Universitaire Pers.

Kersloot, J., & Kauko, T. (2004). Measurement of housing preferences – A comparison of research activity in the Netherlands and Finland. *Nordic Journal of Surveying and Real Estate Research, 1*, 144–163.

Lichtenstein, S., & Slovic, P. (Eds.). (2006). *The construction of preference*. New York: Cambridge University.

Lindberg, E., Garling, T., & Montgomery, H. (1989). Belief-value structures as determinants of consumer-behavior – A study of housing preferences and choices. *Journal of Consumer Policy, 12*(2), 119–137.

Maclennan, D. (1977). Information, space and measurement of housing preferences and demand. *Scottish Journal of Political Economy, 24*, 97–115.

Molin, E., Oppewal, H., & Timmermans, H. (1996). Predicting consumer response to new housing: A stated choice experiment. *Netherlands Journal of Housing and the Built Environment, 11*(3), 297–311.

Molin, E. J. E., Oppewal, H., & Timmermans, H. J. P. (1997). Modeling group preferences using a decompositional preference approach. *Group Decision and Negotiation, 6*, 339–350.

Molin, E., Oppewal, H., & Timmermans, H. (1999). Group-based versus individual-based conjoint preference models of residential preferences: A comparative test. *Environment and Planning A, 31*(11), 1935–1947.

Molin, E. J. E., Oppewal, H., & Timmermans, H. J. P. (2002). Conjoint modeling of residential group preferences: A comparison of the internal validity of hierarchical information integration approaches. *Journal of Geographical System, 4*, 343–358.

Musterd, S. (Ed.). (1989). *Methoden voor woning- en woonmilieubehoefte onderzoek*. Amsterdam: SISWO.

Nevin, B., Lee, P., Goodson, L., Groves, R., Hall, S., Murie, A., & Phillimore, J. (2001). *Liverpool's housing market research programme 1999/2001: A review of the main findings and policy recommendations*. Birmingham: University of Birmingham, Centre for Urban and Regional Studies.

Payne, J. W., Bettman, J. R., & Schkade, D. A. (1999). Measuring constructed preferences: Towards a building code. *Journal of Risk and Uncertainty, 19*(1–3), 243–270.

Timmermans, H., Molin, E. J. E., & van Noortwijk, L. (1994). Housing choice processes: Stated versus revealed modelling approaches. *Journal of Housing and the Built Environment, 9*(3), 215–227.

Veldhuisen, K. J., & Timmermans, H. J. P. (1984). Specification of individual residential utility-functions – A comparative-analysis of 3 measurement procedures. *Environment and Planning A, 16*(12), 1573–1582.

Vriens, M. (1997). Het meten van preferentie structuren. In *Jaarboek van de Nederlandse Vereniging voor Marktonderzoek en Informatiemanagement* (pp. 241–258). Haarlem: De Vrieseborch.

Vromraad (2009). Wonen in ruimte en tijd. *Een zoektocht naar sociaal-culturele trends in het wonen*. Advies 072.

Walker, B., Marsh, A., Wardman, M., & Niner, P. (2002). Modelling tenants' choices in the public rented sector: A stated preference approach. *Urban Studies, 39*, 665–688.

About the Authors

We kindly invite you to send your additional information, remarks and corrections to Sylvia Jansen at s.j.t.jansen@tudelft.nl

H.J.F.M. (Harry) Boumeester, PhD, is working at the OTB Research Institute for the Built Environment, which is part of Delft University of Technology. He specializes in housing market analysis and research on housing demand and housing preference.
h.j.f.m.boumeester@tudelft.nl

H.C.C.H. (Henny) Coolen, PhD, is working at the OTB Research Institute for the Built Environment, which is part of Delft University of Technology. He specializes in housing preference research, the meaning of dwellings, and in quantitative and qualitative research methods.
h.c.c.h.coolen@tudelft.nl

R.W. (Roland) Goetgeluk, PhD, was during the realization of this book working at the OTB Research Institute for the Built Environment, which is part of Delft University of Technology. He is now working at ABF Research Delft, The Netherlands. He specializes in migration, residential mobility, housing policy studies and has a special interest in housing market and land-use simulation models.
Roland.Goetgeluk@abf.nl

C. (Carola) de Groot, MSc, is working at the Netherlands Environmental Assessment Agency (PBL), the national institute for strategic policy analysis in the field of environment, nature and spatial planning. She specializes in housing research and in particular in the discrepancy between stated mobility intentions and actual mobility behavior at the micro level using longitudinal analysis methods.
carola.degroot@pbl.nl

S.J.T. (Sylvia) Jansen, PhD, is working at the OTB Research Institute for the Built Environment, which is part of Delft University of Technology. She specializes in the methodology of housing preference research and has a special interest in the multi-attribute utility method and in the relationship between lifestyle and housing.
s.j.t.jansen@tudelft.nl

S.J.T. Jansen et al. (eds.), *The Measurement and Analysis of Housing Preference and Choice*, DOI 10.1007/978-90-481-8894-9,

M.J. (Marnix) Koopman, MSc, was during the realization of this book working at the OTB Research Institute for the Built Environment, which is part of Delft University of Technology. He is now working at the Netherlands Environmental Assessment Agency (PBL), the national institute for strategic policy analysis in the field of environment, nature and spatial planning. He specializes in urban economics, housing market modeling and performance measurement.
m.j.koopman@tudelft.nl

E.J.E. (Eric) Molin, PhD, is working at the Faculty of Technology, Policy and Management, which is part of Delft University of Technology. His current research interest involves behavioral analysis and choice modeling in Transport, especially with respect to multimodal transport systems and new transport technology. In addition, he is interested in stated preference experiments (conjoint analysis), particularly in advanced experiments such as hierarchical information integration, context-dependent and group experiments, which, among others, he applied to model residential preferences.
e.j.e.molin@tudelft.nl

J.P.J. (Jeroen) Singelenberg, MSc, is working at SEV, national agency for experiments in housing, Rotterdam; he specializes in housing for older and handicapped people, homeless people and migrant workers and in water-related housing.
singelenberg@sev.nl

Index